D0909340

Advances in Teaching Physical Chemistry

ACS SYMPOSIUM SERIES **973**

Advances in Teaching Physical Chemistry

Mark D. Ellison, Editor
Ursinus College

Tracy A. Schoolcraft, Editor
Shippensburg University

Sponsored by the
Divisions of Chemical Education, Inc. and
Physical Chemistry

American Chemical Society, Washington, DC

Library of Congress Cataloging-in-Publication Data

Advances in teaching physical chemistry / Mark D. Ellison, editor, Tracy A. Schoolcraft, editor ; sponsored by the Divisions of Chemical Education, Inc. and Physical Chemistry.

    p. cm.—(ACS symposium series ; 973)

    Includes bibliographical references and index.

    ISBN 978–0–8412–3998–2 (alk. paper)

    1. Chemistry, Physical and theoretical—Study and teaching.

    I. Ellison, Mark D. (Mark David), 1968- II. Schoolcraft, Tracy A.

QD455.5.A38   2007
541.071′1—dc22

                  2007060790

The paper used in this publication meets the minimum requirements of American National Standard for Information Sciences—Permanence of Paper for Printed Library Materials, ANSI Z39.48–1984.

PRINTED IN THE UNITED STATES OF AMERICA

Foreword

The ACS Symposium Series was first published in 1974 to provide a mechanism for publishing symposia quickly in book form. The purpose of the series is to publish timely, comprehensive books developed from ACS sponsored symposia based on current scientific research. Occasionally, books are developed from symposia sponsored by other organizations when the topic is of keen interest to the chemistry audience.

Before agreeing to publish a book, the proposed table of contents is reviewed for appropriate and comprehensive coverage and for interest to the audience. Some papers may be excluded to better focus the book; others may be added to provide comprehensiveness. When appropriate, overview or introductory chapters are added. Drafts of chapters are peer-reviewed prior to final acceptance or rejection, and manuscripts are prepared in camera-ready format.

As a rule, only original research papers and original review papers are included in the volumes. Verbatim reproductions of previously published papers are not accepted.

ACS Books Department

Contents

Laboratory Literature Review

Problem-Solving Issues in Quantum Mechanics

Using Computers to Aid in the Teaching of Physical Chemistry

American Chemical Society Examinations

Innovative Ways of Teaching

Indexes

Advances in Teaching Physical Chemistry

Chapter 1

Advances in Teaching Physical Chemistry: Overview

Mark D. Ellison[1] and Tracy A. Schoolcraft[2]

[1]Department of Chemistry, Ursinus College, Collegeville, PA 19426
[2]Department of Chemistry, Shippensburg University, 1817 Old Main Drive, Shippensburg, PA 17257

Many physical chemists think that the teaching of physical chemistry is currently experiencing a crisis. First, most students enter a physical chemistry course less prepared, particularly in math, than they did in years past. Second, teachers of physical chemistry face numerous vexing questions, including: What do we want to accomplish with the physical chemistry course? What topics we present to a diverse audience? How much of each subject should we emphasize? How can we be certain that the students master the material?

More than a dozen years ago, *Physical Chemistry: Developing A Dynamic Curriculum* was published. (*1*) This landmark book was a welcome advance in the quest to modernize the physical chemistry curriculum. It arose from a Pew Charitable Trust project and NSF Undergraduate Faculty Enhancement workshops at the University of Northern Colorado. These workshops were well-attended and provided clear evidence of the strong need for reform in the physical chemistry curriculum at that time. *Developing a Dynamic Curriculum* included advances in classroom content, laboratory exercises, and writing in physical chemistry, although the bulk of the book is devoted to specific laboratory experiments. Several symposia on aspects of the physical chemistry curriculum have been conducted at ACS and other meetings since 1993. However, no formal assessment of the progress made or the need for further changes has been provided. The time is ripe to bring the community up to date.

Since 1993, much has changed in physical chemistry. The field is dynamic and growing, which adds to the difficulty of teaching the subject. If new material is added to convey to the students the excitement of new discoveries, what old material will be cut from the course? The richness of the field means that difficult choices need to be made. However, we think that including recent discoveries in the course material is essential and that three developments in that time span serve as nice examples to illustrate the future of teaching physical chemistry: Ahmed Zewail has elucidated chemical reactions at the femtosecond timescale, (2) a complete solution for ionization of a hydrogen atom by electron scattering has been calculated, (3) and scientists have designed nanoparticles with tunable optical and electrical properties. (4)

Zewail's work demonstrates that physical chemistry can probe the most fundamental details of chemical reactions. What could be more central to the science of chemistry than understanding each step in a chemical reaction?

The finding of an exact solution for the scattering of an electron from a hydrogen atom exemplifies the current power of computational chemistry. In the mid-1990's, specialized workstations were necessary to carry out calculations, and an ab initio calculation on even a small molecule could take an entire afternoon. Now, a desktop computer can run complex calculations in minutes. All students have at their fingertips the means to explore structure–function relationships, or to construct sophisticated models of chemical systems. In the coming decades, computational chemistry will become an integral part of most chemists' work, so our students must learn how to use computational methods and how to determine which ones are appropriate for their applications.

Finally, "nano" has become a buzzword of the times. As several authors in this book point out, nanoscience is fundamentally chemistry. The design and synthesis of nanoscale objects is based almost entirely on the principles of chemistry, meaning that the study of the synthesis and properties of these systems is well within the domain of physical chemistry. The potential of this rapidly developing field has captured the imagination of the public and, we hope, our students. Nanoscience can allow us to show students that physical chemistry is interesting, highly useful, and—dare we say—fun!

These three examples show the tremendous power of physical chemistry, and helping our students to develop an understanding of these advances drives us to use examples from cutting-edge chemistry in our physical chemistry courses. However, in a talk at a recent ACS National Meeting, Professor Richard N. Zare of Stanford University cautioned, "In terms of incorporating material from present-day research, I would emphasize how physical chemistry lays a basis for understanding processes in living systems. I am mindful that if I stress too much the most recent developments, I would hinder the learning of fundamentals that are of timeless value." (5)

Clearly, the teaching of physical chemistry remains a challenging prospect. Most physical chemistry instructors would agree that we want our students to gain certain proficiencies from our courses. These include sound knowledge of the fundamentals of physical chemistry, increased ability in logical reasoning, improved critical thinking skills, the ability to model chemical systems, and facility in analyzing data. We can certainly gain from sharing successful and not-so-successful efforts to achieve these goals. In that spirit, we have assembled this book, which arises from a symposium at the 230th National Meeting of the ACS in Washington, DC in August 2005: *Physical Chemistry Curriculum Reform: Where Are We Now and Where Are We Going?* This symposium brought together a wide variety of speakers from government, industry, and academe. Many of these speakers agreed to write chapters based on their talks so that we could bring you a sense of the current state of affairs in physical chemistry education.

As Gerald van Hecke and Robert Mortimer explain in their chapters, instructors of physical chemistry must make numerous choices when designing their courses. Some of the important questions they cite include:

- Who are the students in the course, what is their background, and what are their goals?
- What content will I teach?
- From what standpoint, macroscopic or microscopic, will I approach the material?
- What instructional methods will I use?
- How will I handle the laboratory portion of the course?
- What sort of assignments will I give?
- How will I assess student learning?
- How will I use computers and multimedia technology to facilitate student learning?

Obviously, there are many "right" answers (and certainly some wrong answers) to these questions. To begin with, in a recent article, Zielinski and Schwenz suggested that instructors should focus on teaching a core group of subjects well. (6) That theme is also apparent in many of the chapters in the book. Some authors, such as Robert Mortimer and Arthur Ellis, also make this point and offer some general suggestions for the reader to consider. Other authors, including Gerald van Hecke and Peter Atkins, go further and provide specific recommendations for the content they believe should be covered. It is interesting that all of these authors, particularly Ellis, note the rising importance of nanoscience as a subject of instruction in physical chemistry.

Since 1993, a tremendous amount of work has been done to study and evaluate methods of teaching physical chemistry. The various modes of learning and constructing knowledge as applied to physical chemistry have been summarized by Zielinski and Schwenz. (6) A teacher of physical chemistry who wishes to be effective should certainly be familiar with the literature concerning chemical education, particularly as it applies to physical chemistry. A very extensive review of this literature, expanding on the results of Zielinski and Schwenz, is provided in the chapter by Georgios Tsaparlis.

While one is pondering the literature on physical chemistry education, it is worthwhile to think about the philosophy of chemistry as well. Eric Scerri presents a discussion of the implications of the philosophy of science on the teaching of physical chemistry. Often, physical chemistry is viewed as a means of reducing chemical behavior to physics. Eric Scerri points out the pros and cons of this viewpoint and their implications for chemistry education.

Student learning can be enhanced by several factors, including a demonstration of relevance to current research and improved methods of communicating the abstract concepts in physical chemistry. For instance, numerous examples in physical chemistry come from outdated experiments. Michelle Francl reports on her project to overcome this drawback by teaching students the fundamentals of physical chemistry using recent articles from the literature.

To understand the basic steps of a chemical reaction a la Zewail or to understand the properties of nanoparticles, students must grasp the fundamentals of quantum mechanics. However, most students find this subject very difficult. David Gardner and George Bodner present a summary of their research on the teaching and learning of quantum mechanics at the undergraduate level. They conclude that students in quantum mechanics courses develop (or have developed by the time they enter the course) a mindset for performing numerical calculations. However, this mindset, they have found, comes at the expense of understanding concepts.

Because computational chemistry models require critical thought about the system being modeled, the use of such models in the undergraduate chemistry curriculum can have strongly positive results. Exercises in computational chemistry can help students focus on the chemical concepts and the modeling of chemical systems, rather than the details of the mathematical operations. In her chapter, Theresa Zielinski describes the use of Symbolic Mathematical Engines (SMEs) in physical chemistry. SMEs perform the mathematical processesing, allowing students to focus on understanding the underlying concepts. Using SME documents available at Theresa Zielinski's web page (7) or the Journal of Chemical Education SymMath web page (8) in your physical chemistry course might help alleviate the counterproductive problem-solving mindset found by Gardner and Bodner.

Ideas for surmounting this challenge can be found in several other chapters in this book as well. Chrystal Bruce, Carribeth L. Bliem, and John M. Papanikolas describe a free software program that can be used for guided-inquiry exercises into the behavior of gases, both real and ideal, and liquids. In their chapter, they describe open-ended assignments in which students must critically analyze the "results" of the simulation and compare them to experimental data. This approach nicely couples the mathematical formalism with the underlying molecular behavior. Similarly, Jurgen Schnitker describes a commercial software program that can model many more systems. As he notes, the simulation interface allows students to "experiment" and observe the effects of changing variables on their own, moving them away from "cookbook" exercises and permitting them to explore the relationships between state variables in different systems. Finally, Roseanne Sension follows with a description of a course at the University of Michigan that directly involves students in constructing and applying computer models to chemical systems.

This course emphasizes the thought process of building a model and also searching for situations in which the model fails or is inadequate. Clearly, these are important skills for our students to develop.

We point the reader to numerous sources to find information on using molecular mechanics and ab initio calculations in the physical chemistry curriculum. First, Warren Hehre, who presented a talk at the symposium but did not author a chapter for this book, has written a comprehensive description of molecular mechanics and ab initio calculations. (9) An example of using computational chemistry to understand the role of chlorine oxides in stratospheric chemistry can be found in the *Journal of Chemical Education.* (*10*) Also, several workbooks are available with computational chemistry exercises for students to carry out. (*11-13*)

Computational chemistry offers many advantages to teachers of physical chemistry. It can help students learn the material and develop critical thinking skills. As noted before, most students will probably use some sort of computational method in their chemistry careers, so it provides students with important experience. Furthermore, computational chemistry is much more accessible to undergraduate students than it was a decade ago. Desktop computers now have sufficient resources to calculate the properties of illustrative and interesting chemical systems. Computational software is also becoming more affordable. Students can now use computers to help them visualize and understand many aspects of physical chemistry. However, physical chemistry is also an experimental science, and computational models are still judged against experimental results.

Emphasizing experimental physical chemistry is still incredibly important. Zewail's experiments opened the door to understanding chemical reaction dynamics on an entirely new level. In addition, theoretical understanding of nanoscience currently lags behind experimental work in some areas. Thus, it is necessary for students to receive a strong background in experimental physical chemistry. Given that it is now often easier to design and implement computational exercises than laboratory experiments, there is a danger that experimental physical chemistry could suffer. As one speaker at the symposium noted to one of us (MDE), he hoped that experimental physical chemistry would not be displaced by computational chemistry.

Choosing laboratory experiments is always a difficult task for a physical chemistry instructor. One must balance the desire to have students complete exercises that demonstrate modern physical chemistry against the reality of limited resources. Although femtosecond lasers will probably never be found in the undergraduate teaching laboratory, it is still possible to have students use modern instrumentation. Sam Abrash has completed an excellent, thorough, and exhaustive review of physical chemistry laboratory experiments published since 1993. In his chapter, you should be able to find exercises suitable for your pedagogical goals and available resources.

Once you have chosen course content, you will want to think about how you will spend classroom time. Many professors use the traditional lecture, sometimes supplemented with small-group work. However, in the past decade, there has been a growing movement toward active-learning strategies in the classroom. One of these methods is described in a chapter from Jim Spencer and Rick Moog about using Process-Oriented Guided-Inquiry Learning (POGIL) to actively involve students in the physical chemistry classroom. Their chapter contains an overview of the technique and an example of using it in the classroom; more details can be found at the POGIL web site. (*14*) Other guided-inquiry processes, such as Physical Chemistry On Line (PCOL) (*15*) and peer-led team learning (PLTL) (*16*) are described elsewhere.

Assessing student knowledge is tremendously important to teachers of physical chemistry. Richard Schwenz explains the evolution of the ACS exam in physical chemistry and provides insight into the process by which questions are written. Whether or not you use the ACS exam in your courses, this chapter will help you formulate better questions of your own.

Not all of the chapters in this book fit neatly into categories, but they are related to the main themes. Creighton University, as explained by HollyAnn Harris, has added a course on mathematics for physical chemistry and condensed the traditional two-semester physical chemistry sequence down to a single semester. This change has not led to decreased student learning, and it does allow for an increased focus on the concepts of physical chemistry in the course. Finally, Jim LoBue and Brian Koehler present an interesting argument for reversing the traditional order of topics in physical chemistry and discussing kinetics first. This approach has the advantage of starting with topics that the students have covered previously in general chemistry courses and then gradually increasing the difficulty of the mathematics.

We hope that this book will serve as a "road map to the future" of physical chemistry education. It is a road map in the sense that there are many paths, most of which lead to the same destination. For instance, you might choose to help your students develop critical thinking skills by using SME exercises in your course, or you might have students learn the fundamentals using recent, relevant literature papers. The path you choose will depend on your background and strengths, your students, and the resources at your disposal. There is no one "right" answer, but we sincerely hope that reading this book will help you chart successfully your own course. You do not need to be encyclopedic and cover every topic. Rather, you just need to provide the students with the fundamentals and inspire them to learn on their own. So, where are we now and where are we going? Boldly into the future with more tools at our disposal and more knowledge of how our students learn.

References

1. Schwenz, R. W.; Moore, R. J. *Physical Chemistry: Developing a Dynamic Curriculum*; American Chemical Society: Washington, DC, 1993.

2. See, for example: Zewail, A. H. *J. Phys. Chem.* **1996**, *100*, 12701.
3. Rescigno, T. N.; Baertschy, M.; Isaacs, W. A.; McCurdy, C. W. *Science* **1999**, *286*, 2474.
4. See, for example: Bawendi, M. G.; Wilson, W. L.; Rothberg, L.; Carroll, P. J.; Jedju, T. M. et al. *Phys. Rev. Lett.* **1990**, *65*, 1623.
5. Zare, R. N. How I would revise physical chemistry for undergraduates. In *Proceedings of the 229th National Meeting of the American Chemical Society*; San Diego, CA, 2005, American Chemical Society, Washington, DC.
6. Zielinski, T. J.; Schwenz, R. W. *Chemical Educator* **2004**, *9*, 1.
7. Mathcad in Physical Chemistry: http://bluehawk.monmouth.edu/~tzielins/mathcad/Lists/index.htm (accessed July 20, 2006).
8. Journal of Chemical Education Symbolic Mathematics in Chemistry: http://jchemed.chem.wisc.edu/JCEDLib/SymMath/index.html (accessed July 20, 2006).
9. Hehre, W. J. *A Guide to Molecular Mechanics and Quantum Chemical Calculations*; Wavefunction, Inc.: Irvine, CA, 2003.
10. Whisnant, D. M.; Lever, L.; Howe, J. *J. Chem. Educ.* **2005**, *82*, 334.
11. *Teaching with CAChe Instructor Workbook*; Wong, C. and Currie, J., Eds. Available for download at: http://www.cachesoftware.com/cache/education.shtml (accessed July 20, 2006).
12. Caffery, M. L.; Dobosh, P. A.; Richardson, D. M. *Laboratory Exercises Using HyperChem*; Hypercube, Inc.: Gainesville, FL, 1998.
13. Forseman, J. B.; Frisch, A. *Exploring Chemistry with Electronic Structure Methods*, 2nd edition; Gaussian Inc.: Pittsburgh, PA, 1996.
14. Process-Oriented Guided-Inquiry Learning: http://www.pogil.org/ (Accessed July 20, 2006).
15. Towns, M.; Sauder, D.; Whisnant, D.; Zielinski, T. J. *J. Chem. Educ.* **2001**, *78*, 414.
16. Quantum Mechanics: Peer-Led Team-Learning Workshops: http://quantum.bu.edu/PLTL/index.html (accessed July 20, 2006).

Choosing Content
for the Physical Chemistry
Course

Chapter 2

What to Teach in Physical Chemistry: Is There a Single Answer?

Gerald R. Van Hecke

Department of Chemistry, Harvey Mudd College, 301 Platt Boulevard, Claremont, CA 91711

Preview

Few would be surprised with an answer of NO to the title question. Moreover, the answer "no" leaves us with the opportunity for discussion, agreements and disagreements. This article is a collection of impressions/recollections gathered from teaching physical chemistry for more than 35 years. To inform you of the background for my impressions and probable bias, some personal data is included at the end of this article.

Clearly you, the readers, have some background teaching physical chemistry and your background will undoubtedly influence your view of my comments. Do you teach?:

PChem at a PUI?	Other?	
Junior year?	Senior year?	
One semester?	Two semesters?	
Is your order of order of topics?:	Microscopic?	Macroscopic?
Is your student audience:	Chemists?	A mixture of Other?
		chemists/engineers?

Of course I will not know your answer to these questions, but I assume the answers to these questions shape the nature of your course.

My remarks here will focus only on suggested course content and not on modes of presentation such as case studies, guided-inquiry, etc. I will not discuss the physical chemistry laboratory either. Modernizing the physical chemistry laboratory has been the subject of several recent meetings, a published book (*1*), and the chapter by Sam Abrash elsewhere in this book.

My remarks are organized as follows:

- Guides to content
- The irreducible minimum
- An organization scheme for a modern physical chemistry course
- Macroscopic topics
- Microscopic topics
- Summary

Guides to content

The theme of my comments is that whatever your answers to the questions posed above, you cannot teach physical chemistry without taking into account your environment, that is, the context in which you teach. Moreover, you need to take into account your students and their preparation. But today, often overlooked are the constraints of your departmental and institutional goals and the goal for what future you are preparing your students.

How then should the content of physical chemistry be guided by your student audience?

Suppose your audience consisted of chemists only. The concept of a distribution coefficient is a topic that should definitely be discussed but is often not. If you have chemists and engineers, you cannot ignore heat engines; even for chemists alone discussions of heat engines could be omitted. Entropy can be introduced without invoking heat engines. Clearly, enzyme kinetics would be an area to be covered if your audience included biochemists and chemists. If biologists were in your class, a decent treatment of osmotic pressure would need to be presented. The single topics mentioned depending on the audience are minimal suggestions drawn from some larger theme, for example, distribution coefficients are a small but an important portion of phase behavior. The suggested topics are ones that should definitely be discussed depending on your audience but are often neglected. (Chemists and engineers always see a treatment of engines, though.)

Why discuss distribution coefficients? Most everyone is familiar with the demonstration of iodine distributed between an organic and an aqueous layer. However, distribution equilibria are at the heart of many separation processes from liquid-liquid extractions to virtually every type of chromatography in which the distribution of the solute between the mobile phase and the stationary phase determines the effectiveness of the separation. In the practice of analytical chromatography, distribution coefficients are often called partition coefficients but the concept is identical, only the names have changed. The temperature dependence of a distribution coefficient is at the heart of temperature programming in gas-liquid chromatography (GC), and analyses of the temperature behavior depend on the heats of solution of the distributed solutes. Indeed, GC measurements have been used to measure heats of solution.

As engineers still deal with heat flow and heat engines, you cannot avoid discussions of the Carnot cycle and heat engines. However, working fluids other than ideal gases, for example, elastomers, provide an interesting diversion. The Carnot cycle has been worked out for a rubber band, that is, an elastomer as the working substance. All the Carnot cycle conclusions can be derived using a substance that is far more visible (realizable) than an ideal gas. (2)

If you have a class with biochemists, clearly the area of enzyme kinetics is practically mandatory. If biologists are mixed in with the biochemists, osmotic pressure is an important concept to cover carefully and a concept typically not well covered in general chemistry and in most physical chemistry texts or classes. A quick example: what is a 2 Osmolar solution of sodium chloride? Such concentration units are used when dispensing various saline solutions in hospitals. What is the origin of the unit? A 1 M NaCl solution dissociates into two ions that would double the osmotic pressure of a non dissociating solute. Thus, the 1 M solution of NaCl becomes a 2 Osmolar solution. Other examples abound – the bursting pressure of a cell relates to the osmotic pressure of the serum in which the cell finds itself.

Guides to content: current and future preparation

Student preparation is clearly important. The placement of physical chemistry late in the curriculum is most often justified by the argument that the student would be inadequately prepared for the physics/mathematics requirements of the course before their junior or senior year. Personally I find this to be a bogus argument, one that does not give very much credit to the notion that students learn (best?) when motivated by a need to know. Our institution has taught physical chemistry to sophomores for more than 50 years, but more on that later. Obviously, you cannot ignore the students' preparation though you may need to make up for the lack of preparation in your course. If your course comes in the senior year, it is unlikely that any other course will depend on its content. If you teach the course earlier than the senior year, you should be aware of how the content can better inform other courses in the chemistry, perhaps even the biology curriculum.

Guides to content: departmental and institutional goals

The goals of your department must be taken into account. This is especially true if the course is part of a major. Is the department ACS-accredited and is the major path such that the student will be ACS-certified? These standards place restrictions on the extent and content of the physical chemistry course that meets these standards. Moreover, institutions themselves are placing large theme goals on courses within the institution. Many of these institutional objectives seek to acquaint the student with the influence of science on society and the influence of society on science. At Harvey Mudd our mission statement contains the phrase " . . . to understand the impact of their work on society." The intent is clearly present though I have to confess my practice is often too parenthetical and misses making a fuller impact. The second of the two objectives mentioned is all too often altogether overlooked. Consider for a minute how often the direction of science is determined by the funding available from the NSF, NIH, or the pressure of public opinion? To dwell on this point a bit longer, consider that every proposal submitted to the NSF and NIH (and increasing to other funders as well) must include a section describing the broader impact of the proposed work.

The previous points lead to the conclusion that physical chemistry cannot be taught in isolation. Those who teach introductory physical chemistry must take into consideration all of the mentioned factors ultimately to decide the course content. But further, I will make two assertions that are also cautions to all teachers of physical chemistry:

1. The examples used to illustrate concepts need to be drawn from real world situations. Students want (and need) relevance.
2. The wisdom of teaching your own expertise in the course is always questionable.

The first caution is probably intuitive and obvious. The second might be questioned. This caution stems from the assumption made all too often that we can at the drop of a hat speak eloquently on topics of personal great interest and experience. The difficulty is that when we address an audience of beginners, we often forget those little definitions of concepts and phrases that we struggled with on our first exposure to the subject. Assuming our listeners will just naturally understand those tricky parts is probably a very bad assumption. We can teach our expertise but we must remember to begin at the beginning without assuming any knowledge on the part of the listener.

What content should be an irreducible minimum assuming a one-year course?

Now comes perhaps a controversial part of this presentation. Course organizational schemes start from a microscopic view of matter and work toward the macroscopic or vice versa. Today working from the macroscopic to microscopic world view is more common, if for no other reason than that is the way most physical chemistry textbooks are organized. It probably really does not matter which approach is used. I have used both but have settled on the traditional macro to micro approach to reduce chapter jumping in the text of choice. In the scheme below perhaps nothing is too exciting except that the nanoworld is a new player in our world view. The physics and chemistry of nanoparticles clearly seem different and will be a fertile field of study for a long time. This view was promoted at a recent ACS national meeting symposium on nanoscience in physical chemistry. As a parenthetical comment, I would prefer to promote the term nanoscience rather than nanotechnology. (*3*)

An organizational scheme for a modern physical chemistry course

Following the comment on macroscopic then microscopic views made above, a suggestion for the minimum topics to cover would be:

Macroscopic
    Thermodynamics
        Laws, state variables, and mathematics
        Equations of state
        Equilibria: chemical, phase
        Transport properties
    Time-dependent processes
        Chemical kinetics
        Activated processes

Microscopic
> *Statistical thermodynamics*
> *The nanoworld*
> *Quantum world*
>> A box
>> H atom
>> Chemical bonding
> *Interrogation: spectroscopy*

Macroscopic views

Consider the general area of laws, state variables and mathematics. When discussing these topics consider adding some enrichments.

- 1st Law total energy = internal + *kinetic* + *potential energies* of the system
- The *curse of Δ* and thermodynamic changes
- Why only pV work? *Electrical and gravitational work?*
- What is in a name? Gibbs free energy, Free enthalpy, Gibbs energy, *Gibbs potential*
- The dreaded partial derivatives and the bewildering exact differential, *1-forms*

The first law and thermodynamic changes

Introducing the first law of thermodynamics, the engineers do it right for they include the fact the system might have kinetic energy and potential energy besides internal energy. Most of the time our reaction flasks sit quietly on the table and do not move, thus having no kinetic energy, nor do they drop off the table with a potential change. Moving reactant product streams around a chemical plant however calls for consideration of kinetic and potential energy changes. Our pure chemistry majors should realize this but they would never know this from the treatment of the first law in any of today's common physical chemistry texts.

What is a delta (Δ) but an excuse for teachers to be lazy? I call this the curse of delta. Too often delta is allowed to take on a life of its own without the student understanding that it symbolizes a change in some property between two different thermodynamic states. In fact as I write this piece I noted a colleague who referred to Δ**G** as a state function. The state function is G all right but delta just describes a process. How can a struggling student understand this crucial difference when as instructors we either, do not understand the difference ourselves, or continue to be imprecise in our use of language? Instead of delta

X would it not be more informative to write for a change of enthalpy, for example, if i = initial and f = final,

H(final temperature, final pressure, final composition)
  – H(initial temperature, initial pressure, initial composition)

Even shorter   H(T_f, p_f, composition $_f$) – H(T_i, p_i, composition $_i$)

I shall expand this comment further. Let the process be an isobaric heating of a substance whose constant pressure heat capacity is known.

$$\int_{H_{initial}}^{H_{final}} dH = H(T_{final}, p, composition) - H(T_{initial}, p, composition) = \int_{T_{initial}}^{T_{final}} CpdT$$

Clearly, this format takes more time to write, but consider the benefit to the student challenged to calculate an enthalpy change resulting from an isobaric and isoplethic heating. The expression itself tells the student what must be done, the initial and final temperatures must be known and an expression for a change in enthalpy depending only on temperature must be found because pressure and composition are constant. What does ΔH tell the student? Push this example further.

$$H(T_{final}, p, composition) = H(T_{initial}, p, composition) + \int_{T_{initial}}^{T_{final}} CpdT$$

In this format, we can discuss an absolute value of the enthalpy at the new (final) temperature as related to the enthalpy of the initial state and the energy (enthalpy) that had to be added to reach the final state. The fact that the enthalpies of the two states can have absolute values is clear from this formulation. Again, ΔH tells us nothing about absolute values of the state function enthalpy. If the initial state is considered the same, differences in final state enthalpies can be subtracted to obtain new changes in enthalpy. This is how "steam tables" are constructed and used. However, this method must be applied carefully. The initial state enthalpy can never be zero. Even at absolute zero the enthalpy is not zero but consists at a minimum of zero point vibrational energy and electronic energy. Many newer texts are writing state functions as absolute quantities forgetting that there must be a reference state from which the state function is measured.

Work other than pV

Why not introduce electrical work while the ubiquitous pV work is introduced? Electrical work is important and the chapter on electrochemistry in any physical chemistry text depends on it. As chemists we overuse pV work examples that in fact we never really experience in the laboratory. Introduce different types of work in the first law.

The Gibbs potential

The concept of chemical potential was developed by Gibbs as we all accept. Gibbs developed the notion of the chemical potential of an individual species first. Only later did he define the sum of these chemical potentials, essentially molar potentials, as the concept too many still call free energy, a term never used by Gibbs. (*4*) Just as kinetic energy and potential energy are different, so is the Gibbs (and Helmholtz) potential different from the "energy" terms internal and enthalpy. This is not the place to discourse on general work and potential functions, but the Gibbs potential and the Helmholtz potential are truly potentials while enthalpy and internal energy (U) are energies. (*5*). Helmholtz noted that the function called G was the difference between the total energy of the system, the enthalpy H, and the energy unavailable to the system given by the product of temperature and entropy TS. We know that difference as G = H – TS. This difference was clearly then the energy available to the system, or the energy free to cause a change. Actually, Helmholtz argued the basis of the maximum work that could be obtained. (*6*) Well, a potential is an energy available to cause a change. Unfortunately, the free energy view has dominated instead of the potential energy view. Students understand potential energies. We should emphasize the power of the potential point of view to the students. Just as the potential energy of a beaker on a table allows us to predict that it will fall to a lower potential energy if pushed off the table, the Gibbs potential allows us to predict the direction of a thermodynamic change. I emphasize the predictive power of G by using perhaps the most important thermodynamic derivative in the chemist's arsenal of equations

$$\left(\frac{\partial G}{\partial T} \right)_p = -S$$

An understanding of this derivative shows that the change in the Gibbs potential with temperature is the negative of the entropy of the state in question. A deeper insight is revealed noting that because the entropy of a pure substance always increases with temperature, the potential for change, G, decreases. How does free energy help the student appreciate the concept's predictive power? We as teachers of physical chemistry are often our students own enemies by the imprecise language we continue to use.

Mathematical preparation

The calculus phobia for physical chemistry is greatly overrated. However, a new movement in the mathematics community is important for instructors of physical chemistry to understand. Mathematics curricula no longer, at least in their normal courses covering multivariable calculus, discuss what most of us learned as exact, perfect, or complete differentials. Instead the exact differential is discussed as the 1-form of the 0-form of a function of many variables. We should not be surprised at the blank look on our students' faces when we start a discussion with "as you should remember from the calculus, the exact differential is…" because they will not have ever heard the term before. If you have not talked with your mathematics colleagues on the approach they take to teaching multivariable calculus, such a conversation would be timely and should be informative. (7) Modern teaching of functions of several variables does emphasize the geometric character of the relationship between 0- and 1-forms in what we would have discussed as walks on a state function thermodynamic surface described by several variables. Thus, students today should be better prepared to visualize a change in a thermodynamic variable as a movement from one point on a surface to another point. For example, the surface for the Gibbs potential formed by the independent variables temperature and pressure. A change in G going in the temperature direction on this surface, holding pressure constant, that is along a pressure contour, is described by the entropy. Mathematically, the derivative mentioned above describes such a change. Regardless of 1-forms or exact differentials, however, the acceptance by faculty of student calculus/math phobia should be changed. Practically, a partial derivative is a change of only one variable at a time and students can learn this. If thermodynamic changes are related to walks on surfaces, the fear of the difference between common and partial derivatives can be banished. By accepting a student's claim to not understanding mathematics, we really undermine our subject. To be honest, how many of us truly understand the mathematical nuances of partial derivatives; yet, we glibly multiply, divide, and cancel partial derivatives to develop new partial derivatives?

Equations of state

Equations of state (EOS) offer many rich enhancements to the simple pV = nRT ideal gas law. Obviously, EOS were developed to better calculate p, V, and T, values for real gases. The point here is such equations are excellent vehicles with which to introduce the fact that gases cannot be really treated as point spheres without mutual interactions. Perhaps the best demonstration of the existence of intermolecular forces that can also be quantified is the Joule-Thomson experiment. Too often this experiment is not discussed in the physical chemistry course. It should be. The effect could not exist if intermolecular forces were not real. The practical realization of the effect is the liquefaction of gases, nitrogen and oxygen, especially.

Probably the real reason EOS have been painstakingly developed is to calculate p, V, T properties of real gases for practical, that is, industrial purposes. In the usual physical chemistry course calculations for changes in U, H, G, or A using even the van der Waals EOS have been reduced because of the mathematical difficulties integrating the pV term for changes of state. With modern mathematical packages such as Mathematica, MATAB, or MathCad, calculations are virtually routine even if the necessary integration must be done numerically. A conclusion here is to do more meaningful problems.

Again, with available computational packages, the scope of chemical equilibria can be greatly expanded. Phase equilibria are perhaps more important today than they ever was due to advances and the forthcoming necessary advances in materials needed for electronics, high temperature applications, and replacements for existing materials based on nonrenewable resources.

Chemical equilibria

The chemical processing world more often than not depends on understanding the composition of the mixture resulting from many simultaneous reactions. While solving for 10 compositions in a complex mixture used to be very difficult, again the modern mathematical packages, Mathematica, MATLAB, or MathCad, for example, make such calculations possible in real time. Here again the suggestion is to use real world examples to illustrate the importance of chemical equilibria. Solving the necessary equilibrium expressions by minimizing the Gibbs potential of the total with respect to the extent of reaction (or degree of dissociation) of each component is an excellent way to reinforce the meaning of the Gibbs potential. Further, such a calculation illustrates how to find the minima in a multiparameter space, a calculation that can foreshadow finding conformational energy minima of a complex molecule. Another area for introducing multicomponent equilibria, is in the complex world of biological systems. I will do so myself as I become more familiar with such systems. The book by Alberty is a step in the right direction for all of us to become more knowledgeable in applying equilibrium thermodynamics to this important area. (*8*)

Phase equilibria

It is very hard to believe that one currently used physical chemistry textbook states in its preface that the phase rule is a relic of the 19th century and deserves little emphasis today. Shame. The phase rule will still be viable and crucial long after many quantum calculations have come and gone for lack of agreement with experiment. A wonderful example of the use of the phase rule derives from protein crystallization studies. By simply noting the number of phases and melting behavior of the system, a prediction of the purity of the crystallized protein can be made. To ignore the phase rule is to ignore an intellectual triumph of the 19th century. Probably the reason the phase rule is thought to be difficult to apply is that the notion of chemical components as different from chemical species is not understood well enough by users of the rule, including instructors.

Too often, phase diagrams are shortchanged in a physical chemistry course. This too is a shame. Such diagrams are the lifeblood of materials science and increasingly phase diagrams of biological systems aid the understanding of successful formulations for drug delivery systems.

Probably an example and problems derived from the carbon dioxide-blood buffer system in humans should be in every physical chemistry course. What a rich, complex example this is: from Henry's law for the solubility of carbon dioxide in water (blood) to buffer capacity, that is, the rate of change of the law of mass action with proton concentration. The example can be expanded to include nonideal solutions and activities. How many physical chemistry courses use this wonderful and terribly relevant to life example? First-year medical students learn this material.

Why are isotonic drinks useful? Osmotic pressure in living systems is incredibly important; yet how often is the topic dismissed or merely discussed as a means to measure molecular weight of polymers? Why not consider polymers as biological macromolecules and add to the discussion that a balance of osmotic pressure keeps our cells from bursting – which goes back to why the isotonic sport drinks are useful? Relevance in the examples used in our courses is possible.

Time-dependent processes

Time-related processes truly encompass the real world for how many systems are at equilibrium. It is said that if a human system reaches equilibrium, it is dead. Clearly time-dependent phenomena must be discussed in any physical chemistry course.

Transport properties

Transport properties are often given a short treatment or a treatment too theoretical to be very relevant. The notion that molecules move when driven by some type of concentration gradient is a practical and easily grasped approach. The mathematics can be minimized. Perhaps the most important feature of the kinetic theory of gases is the recognition that macroscopic properties such as pressure and temperature can be derived by suitable averages of the properties of individual molecules. This concept is an important precursor to statistical thermodynamics. Moreover, the notion of a distribution function as a general concept can be made quite concrete with molecular speeds. The concept of a distribution function is one of the most fundamental in science and the molecular speeds distribution offers an excellent introduction to this concept. Of course the transport properties of gas viscosity and heat conductivity can be discussed much more extensively, especially when there are chemical engineers in your audience.

Chemical kinetics

The debate between thermodynamics and kinetics is never ending. Is a reaction or process fast enough or thermodynamically stable enough? The debate has no conclusion, but in the process the student should appreciate the questions to be asked and what approaches might be taken to answer the posed questions. Of course, in introducing kinetics, the traditional concerns about reaction order should be raised. The concerns should involve practical examples, however. Enzyme kinetics needs to be introduced and thoroughly discussed. The alternative is for students to see this material in biology courses sans any significant quantitative treatment. Other practical examples that should be featured instead of the tired example of the mechanism of HI production are polymerization reactions. Here synthetic or biological macromolecules can be used for examples.

Perhaps the biggest oversight physical chemists make when discussing kinetics is the neglect of volume effects. To be sure any chemical engineer who would forget the influence of volume even in a constantly stirred tank reactor would not be long in the profession. The volume alone can affect the rate of the reaction. How many physical chemistry courses, or text books, point that out?

Activated processes

With discussion of chemical kinetics goes the energy of activation of chemical reactions. What is too often ignored is that many other physical processes are activated, that is, involve an energy of activation. Phase transitions are often sluggish because of the activation energy to initiate the phase transition. A simple, common example is supercooling. Viscous liquid flow is also an activated process and most fluids flow more readily at higher temperatures because heat bath energy is available to overcome the flow activation energy. Again these topics are very practical and relevant to everyday experiences.

Perhaps a minisummary at this point should be that thermodynamics is too important and too practical to be cast aside or reduced in pursuit of bio X or quantum Y topics. The often used quotation of Einstein comes to mind here.

"A theory is the more impressive the greater the simplicity of its premises, the more different kinds of things it relates and the more extended its area of applicability." A. Einstein

Microscopic views

Statistical thermodynamics

When introducing statistical thermodynamics, keep the discussion relevant and reduce the mathematics. The microscopic world deserves its fair shake in any treatment of physical chemistry and probably few would disagree with that sentiment. The question is how much and what? In a first course I have found that the introduction of statistical concepts through counting possible arrangements subject to constraints has generally worked for the students. How many ways can a fixed number of particles distribute their available constant total energy among the available energies? This treatment follows Hill (9). A goal of this effort is ultimately to reach the point where the Sackur-Tetrode equation can be used to calculate the entropy of a gas at an arbitrary temperature and pressure. Adding to the Sackur-Tetrode equation the expressions for the rotational and vibrational contributions to the entropy allows the calculation of gas phase entropies of real molecules for comparison with 3rd law entropies. Students, at least the thoughtful ones, are usually amazed that the macroscopic approach and the microscopic approach converge on the same answer to within experimental error. Here, by the way, just presenting the relevant equations rather than deriving them has tremendous merits. Does anyone really need in a first exposure to the subject a laborious discussion of Sterling's approximation or the method of undetermined multipliers? Perhaps a most interesting but in some ways quite challenging direction to go in discussing statistical thermodynamics is its application to model biological systems. An area that is accessible to students would be models of adsorption applied to enzyme lock and key mechanisms. (10)

The nanoworld

What may be a "brave new world" with which to inform and challenge our students? The nanoworld. This is a real and present challenge to provide the best information possible. What this means is that we need to do our best at the time, to incorporate the principles that seem to dictate the behavior of nanomaterials into our physical chemistry courses. (Notice in this discussion other than right now the word nanotechnology will not be mentioned.) Nanoscience and nanochemistry are phrases we should be using to discuss this area. Here again we as chemists are our own worst enemies by not presenting the subject as fundamentally chemistry. To beginning students we risk obscuring the fact that to pursue studies in this field, chemistry not nanotechnology is a very fruitful, if not the best preparation. Enough. What is this nano field? The state of matter between the length scales of 1 to 100 nm, length scales that encompass from a single molecule to a solid. Macromolecules, whether synthetic or natural, fall in this length scale and become then excellent first exemplars of this world's properties. Atomic clusters of species such as gold are other good examples, particularly in terms of the different behaviors, different melting points, self-assembly, often observed in such systems. Carbon nanotubes have a very high electrical conductivity, and the explanation for this phenomenon lies in some mixture of macro and micro views of matter. Because the theoretical explanation of physical properties on these length scales is still developing, discussing the nanoworld in our physical chemistry courses will by necessity be qualitative for some time but discuss this world we must. (*3*)

The quantum world

Approaching the quantum world, content will likely be introductory. At the introductory level mathematical derivations should be at a minimum. The restrictions and meanings of the results are far more important.

The particle in a box

In introducing the quantum world we should focus on a minimum set of topics rather than trying to cover the entire subject in a first exposure. Quantum is an area that is all too often over-emphasized in beginning physical chemistry courses. Context should be a guide to coverage. How much time is possible for coverage? Will the student have the opportunity to see more advanced treatments? Does your coverage follow an inorganic chemistry course where all too often a major portion of the course introduces the Schrödinger equation, its H-atom solutions, the aufbau principle for the periodic table, and molecular orbital theory? With what level of mathematical sophistication are your students comfortable? In introductory quantum, probably too much emphasis is placed on the derivations. Forget them. Present the results and what they mean. Derivations can come during a more advanced course or through independent study by the interested student. A reasoned approach is to make sure the concepts of quantization and the origins for quantization are introduced. Believe or not, the particle-in-the-box example is probably the best vehicle to introduce these concepts. The particle in a box is still a viable model to introduce quantum effects arising from confining potentials, boundary conditions, quantum numbers, probability density, energy states, zero-point energy, and the correspondence principle. Perhaps no single model can introduce all of these fundamental quantum ideas as well without a considerable mathematical burden.

The H atom and chemical bonding

Where to go from the particle in a box? Largely what occurs next depends on time available. I have found that generally not discussing rotational and vibrational energies but going directly to the hydrogen atom works. After all, chemists deal with molecules not rigid rotors. Rotational and vibrational energy treatments are important for spectroscopy, which means the amount of spectroscopy to be covered may require more time discussing these topics. However, it is still probably better to discuss the quantum mechanics of rotators and vibrators when faced with explaining spectra based on these motions.

Because chemists are interested in molecules, going directly to molecules and molecular orbital theory is not a bad direction. However, this direction needs a stop at the hydrogen atom. Going to the H atom from the particle in a box allows discussion of real systems, and introduces angular momentum, energies subject to a binding potential, the quantum numbers of atomic orbitals, and of course, our model for single electron orbitals. These concepts are necessary to discuss any bonding, from valence bond theory, valence electron pair repulsion theory, and molecular orbitals. The introduction of molecular orbital theory would naturally come next. However, perhaps a heretical notion is immediately to use commercially available programs for calculating energy structures and not spend a great deal of time explaining how they work. Spend time instead with bench mark calculations to show how well the programs agree or do not agree with experiment. Modeling calculations are becoming better and better and virtually commonplace today. Time spent on learning their uses and limits is likely better than some advanced discussion on the whys of density functional or Hartree-Fock theory.

Interrogating quantum: spectroscopy

Interrogating the quantum world is done using photon probes through spectroscopy and light scattering. However, all too frequently time is spent introducing exotic forms of spectroscopy, usually those favored by the instructor, rather than an introductory overview of the connection between the energy levels provided to us by quantum methods and the differences between those energy levels provided to us by spectroscopy. Spectroscopy provides the opportunity to draw together the energy scales of our physical world in a rational picture. Potential energy surfaces offer the means to tie many spectroscopic concepts together. By noting the energy differences between the potential energy surfaces that describe the ground and excited electronic energy states of molecules, and the energy differences between the vibrational and rotational energy levels of molecules, scales of energy can be made much more concrete. The connection between single point molecular orbital calculations and the development of potential energy surfaces should be made explicit here. Use of potential energy surfaces with vibrational, then rotational energies superimposed on these surfaces, can make energy scales seem much more real and pertinent. At this point solutions to Schrödinger's equation for vibrational energy especially become much more understandable from a relevant point of view. Discussions of spectroscopy beyond simple electronic (if there is anything simple about electronic spectra), vibrational, and vibrational-rotational are probably best left to a specialized spectroscopy course. This is especially true of NMR. Fluorescence spectroscopy should be mentioned if for no other reason that it experimentally represents an emission spectroscopic technique rather than an absorption one. How detailed spectroscopy is discussed using term symbols to describe electronic states probably depends on the student's knowledge of group theory.

Summary

My final remarks are a summary and a challenge. The content of the physical chemistry course cannot be determined by the instructor based on instinct or favorite topics. The subject matter needs to be shaped by the constraints:

- The students, their goals and preparation
- Departmental and institutional goals
- Relevance and usefulness
- Preparation for subsequent courses

The course content suggested here should be held to the metric of the practical and relevant. The placement of physical chemistry in the senior year reduces the subject to one to be endured and not learned and does a serious injury to the intellectual development of the student. By placing the course so late in the curriculum, explanations in organic, inorganic, etc. are based on the famous handwave model or the admonition of wait to physical chemistry arguments. Moreover, the late placement of physical chemistry in the curriculum also reduces the possibility of exciting students to study physical chemistry. That is truly a sad circumstance and represents a challenge that instructors of physical chemistry should work to change.

So in the future, where are we going? Hopefully, we can proceed along the lines suggested above and not in the direction of letting math phobias or the presumed needs of bioscientists dictate the curriculum. Moreover, the trend to one semester courses, or specialty courses for the life sciences, needs to be resisted. Physical chemistry is a fundamental discipline that seeks to provide the quantitative explanation for chemical phenomena. It should be treated with respect and taught with care.

References

1. *Physical Chemistry: Developing a Dynamic Curriculum,* R.W. Schwenz, R.J. Moore, Eds., American Chemical Society: Washington, D.C. 1993.
2. Kemp, D.L. *Physical Chemistry,* Marcel Dekkar: New York, 1979, Chapter 3, pp. 143-145.
3. *CHED Symposium: Whither goest physical chemistry? What is new for incorporating into the curriculum,* American Chemistry Society National Meeting, San Diego, March 2005.
4. *The Scientific Papers of J. Willard Gibbs,* Vol. 1, Ox Bow Press: Woodbridge, CT, 1993. pp. 62-96.
5. Barry, R.S; Rice, S.A.; Ross, J. *Physical Chemistry,* Oxford University Press: New York, 2000, Chapter 13, pp. 373-376.
6. Lewis, G.N.; Randall, M. *Thermodynamics, Revised by K.S. Pitzer, L.Brewer, 2nd Ed.,* McGraw -Hill: New York, 1961, Chapter 13, pp. 140-141.
7. Colley, S. J. *Vector Calculus, 3rd Ed.,* PrenticeHall: New York, 2006.
8. Alberty, R.A. *Biochemical Thermodynamics: Applications of Mathematica,* John Wiley: New York, 2006.
9. Hill, T.L. *Introduction to Statistical Thermodynamics,* Addison-Wesley: Reading, 1960, Chapter 1, pp 6-12.
10. Ben-Naim, A. *Statistical Thermodynamics for Chemists and Biochemists,* Plenum Press: New York, 1992.

Chapter 3

Decisions in the Physical Chemistry Course

Robert G. Mortimer

Professor Emeritus, Rhodes College, Memphis, TN 38112

A physical chemistry instructor is faced with a number of decisions: the goals and objectives of the course, the level of presentation, the choice of textbook, what topics to include, the sequence of topics, the balance between fundamentals and applications, the amount of homework to assign, the use of classroom time, and so forth. Of the students in the class, only a small fraction might intend to become physical chemists. The physical chemistry instructor must make his or her decisions in this context.

Decisions, Decisions

The predecessor of this volume appeared in 1993, and covers a variety of topics. (*1*) The present volume also contains various schemes for improving the physical chemistry curriculum, as well as new suggestions for the laboratory portion of the course. There have been other workshops and meetings, including a workshop on curricular developments in the analytical sciences sponsored by the NSF and chaired by Prof. Ted Kuwana of the University of Kansas. (*2*) You, the teacher of physical chemistry must decide how to apply this large amount of information and the physical chemistry knowledge that you already possess. You should make these decisions consciously, based on the situation that you face and on your goals and objectives for the course. This essay is primarily an attempt by a retired professor of physical chemistry to comment on some of the decisions he has made in a career of four decades.

Goals and Objectives

Your first decision involves the ultimate goals that you envision for your physical chemistry class and the specific objectives you want them to reach in order to work toward these goals. These goals relate to the development of your students' mental abilities and their mastery of physical chemistry. The mental discipline involved in learning any subject is valuable to a student, but this discipline cannot be learned in the absence of subject matter. Physical chemistry contains mathematical, theoretical, conceptual and descriptive aspects, and provides a coherent body of knowledge that is well suited to foster this mental discipline. All parts of physical chemistry cannot be included in a two-semester course, and you must decide what to include or exclude. Your decision will be informed by your goals for the course and by the population of your class, but the fact that a particular student will likely never use Raman spectroscopy does not mean that learning about Raman spectroscopy is of no value to him or her. A few decades ago some people might have thought that learning about nuclear magnetic resonance would have no value to a future physician.

A possible set of goals for a physical chemistry course might be that the students would be able to:

- Understand chemical and physical theories
- Apply theories to specific applications
- Organize knowledge into a coherent whole
- Analyze a problem and devise a solution scheme
- Apply mathematics to solving problems
- Compete in a global economy

- Work in a group
- Plan a laboratory experiment

Many of these goals would be the same for a course in any subject, and the students should learn that the value of any course goes beyond mastery of its subject material. B. F. Skinner once wrote, "Education is what survives when what has been learned has been forgotten." (3)

Once you have your goals in mind you need to decide how best to assist your students in achieving these goals. This involves designing specific measurable objectives for the course, which should be chosen so that you can assess the extent to which the students have achieved these objectives during the course. A few possible objectives might be for the students to:

- Be able to apply new descriptive chemical information
- Be able to give organized oral presentations
- Be able to write coherently
- Be able to assemble laboratory apparatuses
- Be able to carry out laboratory manipulations
- Be able to carry out an error analysis

The Population of Your Physical Chemistry Class

Once you have your objectives in mind, you must analyze the population of your physical chemistry class in order to plan the accomplishment of your objectives. Most of your students will not become physical chemists. In a career of four decades, I worked with hundreds of physical chemistry students. Among these former students, I can now identify only a handful of professors of physical chemistry. There are several professors who teach other chemical subjects and a few high-school teachers. Those who work in industry work mostly with organic, analytical, or polymer chemistry. The largest group of my former physical chemistry students is made up of physicians and other health care workers.

You should evaluate your students' preparation for the physical chemistry course. Most people agree that today's students are less well prepared than students of a few decades ago, but the students still come to you with widely varying preparation and abilities both in mathematics and in chemistry. Furthermore, their proposed career paths do not necessarily correlate with their preparation. I found it helpful to pass out a take-home orientation quiz on the first day of class. This quiz contained a few mathematics problems, a few physics and chemistry questions, a request for a list of all of the science and mathematics courses taken, and a request for a statement of the student's proposed career. Analysis of the students' responses could be used in planning the course.

The preparation for physical chemistry laboratory is a little harder to judge. Today's students seem less experienced in working with their hands than students of a few decades ago, except in the area of video games. I have used an initial experiment in which the students were required to carry out three simple measurements after assembling simple apparatuses, a technique that I learned from Ed Bair at Indiana University. Close observation of the students as they carry out such an experiment gives some information about their aptitude. This experiment also provided the opportunity to discuss data reduction, error analysis and report writing at the beginning of the course.

Once you have designed objectives for your course and have determined what you can about the students in the class, you have more decisions to make. These might include the following:

- How much time to spend on the course
- The general approach of the course
- Topics to include or exclude
- The order of topics
- How to use class time
- How to relate to your students
- What kind of homework to assign
- Whether to give frequent quizzes
- How many and what kind of examinations to give
- How to determine grades
- What classroom tools to use

Time

You have other things to do and must decide how much time you can spend on your physical chemistry class. You might have a number of research students working with you who need guidance. You might need to write a new grant proposal. You have other assignments such as student advising, committee work, department administrative work, and so on. However, you should not place your physical chemistry course at the bottom of your list of priorities.

The Approach of the Course

You will have to decide whether to use a mathematical approach or a more conceptual or heuristic approach. The students in the United States of America are now less well prepared in mathematics than were the students of a few decades ago. This provides an incentive to make your physical chemistry course less mathematical and more conceptual, and this might be appropriate for students who do not intend to become physical chemists. However, I believe that mathematics is the most important tool of physical chemistry, and that the ability to use mathematics will make your students more able to compete globally in the "flat world." (4)

You will have to decide what to do if your students' mathematical preparation is inadequate for the approach that you choose. You might decide to discuss mathematical topics with the entire class, as the topics are needed. You might decide to offer some mathematics help sessions outside of the regular class meetings. You might decide to refer students to one of the books that are available. (5) In any event, once you have decided on the level of mathematical sophistication that you want your students to achieve, you must help them to achieve it.

Choice of Topics

Physical chemistry has been described as "everything that physical chemists happen to be interested in." You cannot cover all of it in two semesters, but a typical course might contain the following general topics:

- Descriptive behavior of gases and liquids,
- Thermodynamics and its applications
- Dynamics, including gas kinetic theory, transport processes, and chemical kinetics
- Quantum mechanics, including atomic and molecular structure
- Spectroscopy
- Quantum statistical mechanics

There are other topics that might be considered, such as solid state theory and classical statistical mechanics. You must decide how much time to spend on each of your chosen topics. You can identify subtopics that you might omit or to which you can give only an introduction. One of the difficult decisions involves how much to teach about your own research area. You are obviously excited about this area, and will be tempted to spend too much class time on it. Another difficult decision is how much time to spend on topics of current interest such as nanomaterials and environmental chemistry. Your decisions should be guided by the composition of your class. If the class has a lot of premedical students and biochemistry majors in it, they are probably well served by a thorough treatment of thermodynamics and dynamics, and perhaps less well served by a thorough treatment of quantum mechanics and statistical mechanics. If the class is mostly composed of future chemistry graduate students, quantum mechanics and statistical mechanics are more important.

You must also decide how to approach computational chemistry. Many software packages are now available that allow the user to make sophisticated calculations without a detailed knowledge of the underlying theory. However, it is not productive of learning to let the students use the software as a "black box." You must decide how much of this theory to present and how to present it so that the students can appreciate what their calculations mean.

The Sequence of Topics

There is an accelerating movement to place quantum mechanics and spectroscopy in the first semester, with thermodynamics and kinetics in the second semester. If all students take both semesters, this is not a bad idea, especially for future physical chemists. However, many colleges and universities require biochemistry or chemical biology students to take one semester of physical chemistry, and some of these colleges do not offer a separate one-semester course for these students. If a separate one-semester course cannot be offered, it is probably better for these students to place thermodynamics and dynamics in the first semester. I have seen one college catalog that specifies one semester of the standard physical chemistry course for biochemistry majors. This requirement was probably set up when thermodynamics and dynamics were taught in the first semester. However, the physical chemistry course at this college now has quantum mechanics and spectroscopy in the first semester, leaving the biochemistry students without instruction in thermodynamics and dynamics.

The placement of statistical mechanics in the sequence is another issue. I think that careful treatments of thermodynamics and quantum mechanics should precede the presentation of statistical mechanics. This can be accomplished with thermodynamics in the first semester, quantum mechanics in the second semester, followed by statistical mechanics near the end of the course. If statistical mechanics is taught before thermodynamics or quantum mechanics, you must either provide a brief introduction to some of the concepts of these subjects at the beginning of the treatment or integrate it into the treatment.

Use of Class Time

At one time, most college and university courses were taught as lectures, with students passively taking notes while the professor did all of the talking. Such an approach stimulates little student engagement and provides the instructor with little constructive feedback. There are other approaches that are more productive of learning. In one approach the class is divided into groups in which students work together during class time and are asked to report on the work of the groups. It is hoped that one or more of the students will act as mentors for the other students in the group. Another approach is to use what used to be called the "Socratic method," which means that the teacher and the students carry on a dialogue driven by questions posed by the teacher. If you work out example problems in class, student involvement is increased if the students know that they can be asked to carry out steps in the solution or to figure out what the next step should be. You can also ask the students to work problems in class, either individually or in groups.

Relating to Your Students

There are several possible models for relating to your students: You might be a "boss," a "coach," an "older friend," an "uncle," or a "buddy." As a beginning teacher many years ago, I subscribed to the "boss" model. I wore a coat and tie to class and called the students "Mr. Smith" and "Miss Jones." The students eventually asked me to call them by their given names, and I have tried since then to follow the "coach" model. It is important to build a genuine rapport with the students and let them know that you are genuinely interested in them, collectively and individually. However, it is probably a mistake to adopt the "buddy" model and to encourage the students to call you by your given name and "hang out" with them socially. You will have to decide how best to remain approachable but not be on the students' level. You should provide adequate office hours for individual and group consultations, and you might decide to tell the students that you will talk to them outside of your office hours whenever you are available. The important thing is to let them know that you care about their learning and that you and the students are on the same team.

Classroom Tools

You must decide how to use the classroom tools that are available to you. Not many years ago, the only such tools were chalkboards, overhead projectors, and demonstrations. Many classrooms are now equipped with a computer and an associated projector. Some professors prepare slides that contain everything, including equations, so that they do not write on the chalkboard at all. This "canned" approach could encourage students to be passive and to regard the presentation as proceeding as though the students were not present. A better use of the computer and projector is to show slides to illustrate specific points and to carry out real-time simulations. Graphs can be created with values of parameters that are chosen by the students. Software packages such as Spartan and CAChe are sufficiently rapid that molecular modeling can be carried out in real time, allowing students to choose different molecules and different displays. Old-fashioned demonstrations can also be valuable. Building an electrochemical cell in front of the students can illustrate some electrochemical principles. A toy gyroscope can illustrate some things about angular momentum as it precesses in the earth's gravitational field. The *Journal of Chemical Education* has a regular feature on tested demonstrations, and its website lists a number of books containing demonstrations.

In addition to these tools, many useful tools are available on the internet. There are such things as programs that will solve the Hückel equations on line, (6) provide information about the elements, (7) and so forth. Various computer programs are available from the *Journal of Chemical Education*.

Examinations and Quizzes

Exams and quizzes can be tools to increase student learning as well as a means to evaluate the students' progress. It is probably a good idea to have several examinations during a quarter or semester. The traditional mid-term plus final system probably places too much information on one examination and penalizes a student too much if he or she has a bad day. Some professors omit the lowest examination score from the computation of the grades, but I have never done this, because it might make the students think that they can take a chance and fail to prepare for one examination. You might decide to have a variety of examination types. If your institution has an honor system, you can include take-home examinations, either closed book with a time limit, or open book with no time limit. For an in-class examination, you might want to allow the students to bring one sheet or one half-sheet of equations to the examination. I have had students tell me that after making up such a sheet they didn't need to use it in the examination. One thing that I have done is to write out complete solutions for an examination and to hand them out at the end of the examination. The students are more interested in the correct answers then than at any other time, and almost all of them will look closely at the correct answers. For a small class you might want to try individual oral examinations. The students are nervous prior to an oral examination, but afterwards usually say that it was a valuable learning experience.

Sometimes students complain that a 50-minute class period is too short for a meaningful examination, so it might be possible to schedule an examination in the evening or on the weekend and allow 90 minutes or longer for the examination. This also frees up a class period that otherwise would be used for the examination. You might want to use this class period for a review and problem session prior to the examination. During this period you can pose problems similar to those that will be on the examination, and can watch the students work them on the chalkboard or on paper. I have even scheduled such problem sessions outside of the regular class period, with optional attendance.

You will have to decide whether to give frequent quizzes and how much weight to give quiz scores on the final grades. You might decide to give a quiz at the beginning of the class period to encourage students to read the relevant material prior to coming to class. You might decide to give a quiz at the end of the class period to see whether the material covered that day has been absorbed. You might want to let the students know in advance when a quiz will be given, or you might want to give unannounced "pop-quizzes." If you give quizzes, you should try to determine whether they are valuable to the students. I have had classes in which even the better students didn't bother to prepare for quizzes, thinking that their examination scores would counteract any low quiz scores.

Homework

You must decide how much homework to assign and what kinds of problems to assign. It is probably a good idea to assign a mixture of types of problems. The students might complain about problems that require them to carry out derivations, but these can be valuable. "Plug and chug" problems do have value, but problems that require devising the solution scheme or working out a derivation are probably more valuable. If some of the problems are easier than other problems, the students can do the easy ones first and gain some confidence and momentum. Most textbooks include a good selection of homework problems, but you might also want to make up your own problems, especially if you think that homework solutions from previous years are available to the students. It is a good idea not to assign the same problem two years in a row.

You must decide whether to have the student turn their homework in. If the homework is not turned in, you must convince the students that doing the homework will help them with the examinations. If the homework is turned in, it should be graded. In a small college you probably have no student assistant to assist with the grading. I graded homework laboriously for many years, and finally decided that it was not a good use of my time. After that I put out a complete solution to each set of homework the day before it was due and had each student grade his or her own paper. The students were required to turn their papers in for inspection and recording of the score. This procedure seemed to give the students more benefit from doing the homework than handing it in for grading, because they could see their errors as they graded their papers.

Grading

Most of today's students are very grade-conscious, and you must be careful to have a fair system. If the physical chemistry laboratory is combined with the lecture as a single course, you must decide how much weight to give the laboratory course. If the homework is turned in and graded, you need to decide how much weight to give the homework. If you give four examinations during the term and give a final examination, you might give the homework the same weight as one in-term examination and to give the final examination twice this weight. If you give quizzes, you must decide how much weight to give the quizzes.

Once you decide these things, you must decide how to translate numerical scores into letter grades. A variety of options exist, such as 90/80/70/60 or grading on a curve. Grading on a curve has the unintended effect of penalizing students who work together. A fixed grading scheme has various advantages, especially if you intend to use group work as a pedagogical approach. It is not necessary to adhere strictly to the 90/80/70/60 set of cutoff points. My preference was to give difficult examinations and to have lower cutoffs. I would announce the grade cutoff points at the beginning of the course, but would announce that I could lower the cutoff points at the end of the semester, but would not raise them.

The Physical Chemistry Laboratory

The physical chemistry lecture and laboratory might be parts of a single course. They might be separate courses taught by the same professor or by different professors. The physical chemistry laboratory might also be part of a combined laboratory course involving physical chemistry, analytical chemistry, and inorganic chemistry. In any case, you must decide how to make the laboratory and the lecture course work together to maximize student learning. The selection of the experiments requires difficult decisions. There is value in the traditional experiments that reinforce standard lecture topics, and there is value in experiments that relate to more modern techniques. The experiments that require considerable data reduction seem to be the most valuable. You will have to decide how to schedule the experiments. There is value in having the students carry out an experiment when they are studying the related theory in the lecture and in having all of the students do the same experiment at the same time. To do this, it might be possible to have students schedule instrument time outside of the regularly scheduled physical chemistry laboratory period. If you have a budget for apparatus, you must decide how to use it. It might be possible to buy equipment that enables you to have several groups of students doing the same experiment at the same time.

You will have to decide how to structure the laboratory. There is some value in having students work in pairs or teams of three, but there is also the danger that one of the students will do most of the work. You will have to decide whether to have the students set up the apparatus. I have had students break some expensive apparatuses, but they need to learn to set up an experiment. You will have to decide how many experiments to include. My plan was to schedule two periods for each experiment. The first period was for setting up the apparatus and taking the data. The first part of the second period was used for a group discussion of the data reduction, the error analysis, and the report writing and the rest of the period was used for calculations and report writing. Students seem to find error analysis difficult, but it is valuable to them and you will have to decide how much error analysis to require. You will need to decide what kind of laboratory reports to require. The writing of laboratory reports is an opportunity for students to improve their writing skills. Resources are available to help develop this skill. (8) There is value in having students write reports in the same format as a manuscript for a scientific journal, but this is time consuming and you might choose to require briefer reports.

With the pressure to find enough time for all of the topics in the lecture course, you might decide to move some topics from the lecture course into the laboratory course. For example, you might not include polymer chemistry in the lecture course, but might include a polymer experiment in the laboratory, along with some discussion of polymer chemistry. You might also decide to include computational chemistry and molecular modeling in the laboratory course instead of in the lecture course.

The Choice of a Textbook

There are a number of good choices, but you will not be completely satisfied with any textbook. I was not even completely satisfied when I used a textbook that I had written. You will probably choose a textbook that is compatible with your sequence of topics. There is at least one popular two-semester textbook that begins with quantum mechanics. (9) Most of the one-volume textbooks begin with thermodynamics but can accommodate different sequences. There are now physical chemistry textbooks that come in two or even four volumes, which provides for flexibility. In making your choice of textbook, you should consider clarity of presentation for the student. Because you are already familiar with the subject, this can be hard for you to judge. I once chose a textbook that seemed perfectly clear to me, but was not at all clear to the students. Next, you should consider the approach of the book. If you want to teach a more mathematically based course, you will probably decide to choose a textbook that uses this approach and not simply plan to provide supplementary information in class.

Summary

You have a lot of decisions to make in teaching physical chemistry. The most important decisions involve achieving the course objectives and fitting the course to the students' needs. Probably the main thing to remember is to let the students know that you care about them and their education.

References

1. Richard W. Schwenz and Robert J. Moore, *Physical Chemistry – Developing a Dynamic Curriculum*, American Chemical Society, Washington, D. C., 1993.
2. Curricular Developments in the Analytical Sciences, URL http://www.chem.ku.edu/Tkuwana/CurricularDevelopment
3. *New Scientist*, May 1964.
4. Thomas L. Friedman, *The World Is Flat*, Farrar, Straus, and Giroux, New York, 2005, gives a sobering description of the level playing field the world is becoming. Friedman points out that India produces over one hundred thousand graduates in science and engineering each year, and that these graduates have received a very good technical education.
5. James R. Barrante, *Applied Mathematics for Physical Chemistry*, 3rd ed., Pearson Prentice Hall, Upper Saddle River, NJ, 2004, and Robert G. Mortimer, *Mathematics for Physical Chemistry*, 3rd ed., Elsevier Academic Press, San Diego, CA 2005
6. R. M. Hanson,. *J. Chem. Educ.*, 79, 1379 (2002).
7. WebElements, URL http://www.webelements.com
8. M. Robinson, *Write Like a Chemist*, URL http://www4.nau.edu/chemwrite
9. D. A. McQuarrie and J. D. Simon, *Physical Chemistry, A Molecular Approach*, University Science Books, 1997.

Chapter 4

Integrating Research and Education to Create a Dynamic Physical Chemistry Curriculum

Arthur B. Ellis[1,2]

[1]Department of Chemistry, University of Wisconsin at Madison,
1101 University Avenue, Madison, WI 53706
[2]The author is on detail to the National Science Foundation
from the University of Wisconsin at Madison through June 2006

The physical chemistry curriculum can be continuously updated by incorporating results from cutting-edge research and technology into undergraduate classrooms and laboratories. Strategies that facilitate this integration of research and education are discussed and illustrated using examples drawn from nanoscale science and engineering. National Science Foundation programs that support such efforts are described.

A general objective for our educational enterprise should be to imbue it with the same vitality that we take for granted in our research enterprise. This can be accomplished by continuously taking the fruits of research and technology and moving them into the curriculum. As a foundation course for preparing the future technical workforce in the chemical sciences, physical chemistry is well positioned to provide leadership for creating such a dynamic coupling of research and education. The emerging multidisciplinary field of nanoscale science and engineering provides a compelling platform for developing new paradigms for a dynamic physical chemistry curriculum.

The National Nanotechnology Initiative was launched in 2000 by President Clinton and followed in 2003 by the 21st Century Nanotechnology R&D Act that was signed into law by President Bush (*1*). Our planned federal investment of approximately four billion dollars over a four-year period is not unique. Similar investments are being made worldwide and reflect a global awareness that nanotechnology holds tremendous promise as the basis for future technological developments.

Much of the excitement over nanotechnology lies in the new science that is being discovered and the new tools that allow us to manipulate matter at the nanoscale, including individual atoms. A website has been established through a National Science Foundation-supported Materials Research Science and Engineering Center (MRSEC) at the University of Wisconsin-Madison to help move these ideas into educational and outreach venues (2). A video laboratory manual is available on the website that provides step-by-step instructions in the form of videoclips.

In the nanoscale regime, the role of defects becomes more pronounced, quantum effects can dominate systems, and transitions to bulk and ensemble-averaged properties can be investigated. These represent important concepts that can readily be incorporated into the physical chemistry curriculum. For example, nanoscale samples of gold represent a multitude of new allotropes of this element. A simple synthesis of gold nanoparticles is available for use as a physical chemistry laboratory experiment and illustrates the striking change in color for this element when only nanoscale clusters of gold atoms are present (3). Once prepared, these samples can be used to explore optical polarization effects. It is noteworthy that there are substantial synthetic challenges: given that the properties of gold nanoparticles depend markedly on their size and shape, we do not know yet how to create gold nanoparticles of arbitrary dimensions in high yield. More research is needed and these experiments might be extended to explore synthesis-structure-property relationships.

Similar educational opportunities abound for carbon. The diamond and graphite allotropes of carbon have been mainstays of chemistry classes for generations of students and provide a contrast between a three-dimensional structure of great hardness and a two-dimensional structure with lubricant properties, respectively. We now have what can be regarded as zero- and one-dimensional counterparts - buckyballs and carbon nanotubes, respectively - with their rich diversity of structural relatives and physicochemical properties (4). These materials are being employed in a variety of nanoscale devices because of their unusual chemical, mechanical and electrical properties.

Interesting physicochemical properties of nanoscale materials are not restricted to chemical elements. It is easy to prepare nanoparticles of magnetite by combining ferrous and ferric chloride, ammonia, and water (5). Addition of a surfactant produces the visually striking ferrofluid: a magnet placed beneath a puddle of ferrofluid produces remarkable spikes. Technological applications of ferrofluids include making seals for high-speed computer disc drives. Quantum dots of CdSe can also be prepared in a physical chemistry laboratory (6). These dots emit colors across the visible spectrum based on their size and provide an engaging introduction to both nanobiotechnology and to the "particle in a box," a staple of the physical chemistry curriculum.

Light emitting diodes (LEDs) provide extant commercial examples of nanotechnology that can be introduced into the physical chemistry curriculum. LEDs are revolutionizing lighting and display technologies (7). The semiconductors comprising these devices can be grown virtually an atomic layer at a time to create particle-in-a-box quantum structures that produce light at high efficiency at wavelengths determined by the dimensions of the box. Another means for controlling the color of the light is to prepare solid solutions using the periodic table as a design tool.

Tools that are emblematic of nanotechnology are the scanning probe microscopes (8). Their ability to image individual atoms and to position them has opened entirely new vistas in nanoscale science and engineering. Use of scanning probe microscopes to create nanoscale architecture like the "quantum corral" has directly revealed the wavelike behavior of matter at this scale (9). Many of these images are so striking that they provide "teachable moments," prompting students to ask how such an image could have been constructed.

Important objectives of the National Science Foundation (NSF) are to promote the integration of research and education and to help prepare the future technical workforce. Nanotechnology has the potential to attract a diverse, talented group of students to technical careers in much the way that space exploration inspired students of an earlier generation to become scientists and engineers. The foundation has supported the movement of nanoscale science and engineering into the curriculum through awards made under its Nanoscale Science and Engineering Education (NSEE) initiative. Through the Nanotechnology in Undergraduate Education (NUE) program, which is part of the NSEE inititiave, NSF has supported the development of college courses in nanotechnology, new examples of nanotechnology that can be used in existing courses, acquisition of instrumentation and development of software for undergraduate exposure to nanotechnology. A complete listing of NUE awards for fiscal years 2003 through 2005 is available on the NSF website (10). One of the NUE awards has been to the ACS Exams Institute to support the development of standardized test questions that instructors can use if they include nanotechnology in their college courses.

Nanotechnology is a rapidly moving field. In thinking about how to couple the research and education enterprises more dynamically, one possible mechanism is graduate education. Some chemistry graduate students are writing papers and thesis chapters that describe new curricular materials based on current research. One graduate student, for example, who had published several original research papers on nanowires, developed a college laboratory experiment for growing nickel nanowires using inexpensive materials, assisted by an undergraduate. She published this work in the *Journal of Chemical Education* and included it in her thesis as one of the chapters (11, 12). This model could be generalized to allow many more interested students to participate in the creation of instructional materials based on their research interests.

Ultimately, the integration of research and education is a community responsibility that can benefit from broad participation of faculty and co-workers at a variety of stages of professional development. With its broad and fundamental sweep, physical chemistry is an excellent platform for such an effort. The inclusion of examples from other disciplines and multidisciplinary fields like nanotechnology can enrich the physical chemistry curriculum and keep it perennially fresh and exciting for both instructors and students.

Acknowledgments

This material is based upon work by Arthur B. Ellis, conducted while serving at the National Science Foundation and supported by the National Science Foundation. Any opinion, findings, and conclusions or recommendations expressed in this material are those of the author and do not necessarily reflect the views of the National Science Foundation. Funding for the development of instructional materials that involved the author's participation was provided by the National Science Foundation through a Materials Research Science and Engineering Center on Nanostructured Materials and Interfaces (DMR-0079983).

References

1. General information about the National Nanotechnology Initiative is available on the web at URL http://www.nano.gov .
2. Educational resources for nanotechnology are available on the web at URL http://www.mrsec.wisc.edu/nano .
3. URL http://mrsec.wisc.edu/Edetc/nanolab/gold/index.html .
4. URL http://mrsec.wisc.edu/Edetc/cineplex/nanotube/index.html .
5. URL http://mrsec.wisc.edu/Edetc/nanolab/ffexp/index.html .
6. URL http://mrsec.wisc.edu/Edetc/nanolab/CdSe/index.html .
7. Condren, S.M.; Lisensky, G.C.; Ellis, A.B.; Nordell, K.J.; Kuech, T.F.; Stockman, S. *J. Chem. Ed.* **2001**, *78*, 1033-1040.
8. Ellis, A.B.; Geselbracht, M.J.; Johnson, B.J.; Lisensky, G.C.; Robinson, W.R. *Teaching General Chemistry. A Materials Science Companion;* American Chemical Society: Washington, D.C., 1993; pp. 15-23.
9. URL http://www.almaden.ibm.com/vis/stm/corral.html .
10. URL www.nsf.gov and enter NUE into the search engine.
11. Bentley, A.K.; Farhoud, M.; Ellis, A.B.; Lisensky, G.C.; Nickel, A-M.L.; Crone, W.C. *J. Chem. Ed.* **2005**, *82*, 765-768.
12. Bentley, A.K. Ph.D. thesis, University of Wisconsin-Madison, Madison, WI, 2005.

Chapter 5

The Evolution of Physical Chemistry Courses

Peter Atkins

University of Oxford, Lincoln College, Oxford OX1 3DR, United Kingdom

I review the difficulties and opportunities that we need to consider when developing physical chemistry courses. I begin with a comparison of the structure of courses in the USA and the UK, then turn to the question of the order of the course: quantum first or thermodynamics first? I then consider the impact of biology on our courses and then turn to the role of multimedia and graphics. I conclude with an attempt to identify the key equations of physical chemistry.

Introduction

In this chapter, I intend to present the considerations that go into the formulation of physical chemistry courses. They are much the same as go into the formulation of physical chemistry textbooks, of course, for the two modes of delivery go hand in hand. However, I shall do my very best to stand back from my own prejudices and will try to give an even-handed account of a variety of opinions.

Other systems

I shall share the kind of thoughts that go through my head when planning the presentation of physical chemistry. Although my background is in the British system, I have immersed myself for decades in the American system, and will focus on that. However, it may be of interest at the outset to describe very briefly what I perceive as the distinction between the two.

Whereas the American system is horizontal, the British system is vertical. That is, the American system arranges courses in sequence, with what (to be honest) is introductory physical chemistry in the freshman year, then, typically, a physical chemistry course in the junior year. There are modifications of that, of course, but that is the broad picture. By contrast, in the British system, there is not (or at least, until recently, there has not been) a freshman course, on the grounds that high school chemistry is a serious course that in some respects goes beyond an American freshman course. As soon as the college course begins, all three branches are taught in comparable depth and that parallel development continues for all three or four years of the course.

In my view, the British system is unlikely to survive in its pure form, for high schools are finding it difficult, for a whole variety of reasons that are not relevant to this chapter, to bring people up to the standard that such an approach requires. Nevertheless, it is likely that in the UK, the central idea that the branches of chemistry should be developed in parallel is likely to survive.

The problems approaching the American and British systems are quite different in some respects, for American students have already been exposed to quite a lot of .physical chemistry in their freshman year, and are at a more mature stage of their studies by the time they embark on the physical chemistry course proper. They have also, by then, done (but perhaps neither understood nor remembered) more mathematics. In short, they are poised to be more serious than their British counterparts when they start the course. Whether the courses ultimately result in a more competent product is hard to say. It is certainly the case (I think) that after a doctoral program there is little to choose between them in terms of intellectual ability, but the greater emphasis on graduate coursework in the US and the somewhat longer time taken to complete a thesis probably gives the US an edge.

With those general remarks in mind, I shall now focus on the system that prevails in the USA; but those in charge of courses could do worse than to appreciate that there is another side to the Moon that might be invisible to those within its shores.

The order of the course

The first question that comes to mind when considering the structure of a course in physical chemistry is its order. Broadly speaking, there are the camps inhabited by thermodynamics first (T-first) and those inhabited by quantum first (Q-first). I shall set out what I see as the advantages of each approach.

A T-first approach has the intellectual advantage of presenting thermodynamics as an intellectual structure free of any suppositions about the existence and properties of atoms. As such, there is an intellectual purity about it and it is, in a certain sense, model free. There are considerable pedagogical advantages in showing students that a body of knowledge can be constructed on the basis of macroscopic, bulk observations. I am not saying that that is the best way to teach thermodynamics, but it is certainly arguable that they should be exposed to a mode of thought that transcends the microscopic. Of course, they know that there are atoms, and they rightly seek connections to the microscopic world, but by constructing a T-first course they are being exposed to a very special mode of thought.

The second advantage of a T-first approach is that it builds on what students think are familiar concepts, such as work, energy, and temperature. That these are in fact rather sophisticated concepts can be allowed to dawn on them. At least, when they begin the course they are in the seemingly familiar world of macroscopic events and concepts, and there is an easy (but perhaps false) familiarity with the basic ideas of the subject. Moreover, even the mathematics is relatively familiar, with only the concepts of partial and exact derivatives central to its exposition.

The third advantage is that thermodynamics is easy to apply to the problems that chemists commonly encounter and relevant and useful results can be obtained at an early stage without a lot of effort. In a Q-first approach, with its journey through partition functions *en route* to thermodynamics, there is the lingering thought that one ought to be thinking in terms of and using the heavy apparatus of statistical thermodynamics. In other words, the T-first route is immediately and directly applicable without too much clutter; the Q-first route can obscure that essential, pragmatic simplicity.

Then there is the fourth point, the practical issue of timing relevant to other courses and other participants, such as engineers and the correlation of lectures with laboratory. This is a local issue, of course, but it might have a serious impact on the development of courses locally. It seems to me, though, that a chemistry department should establish its own priorities and approach initially, and bend only as others' arguments prove compelling.

And finally, there is a wholly flippant point: that in the T-first approach, the thermodynamics is out of the way and there is something modern and exciting to look forward to, rather than vice versa!

The Q-first approach also has a number of advantages, especially in the context of courses with an American structure where the primary concepts and elementary applications of thermodynamics have already been encountered in the freshman year and, therefore, where some of the language of thermodynamics is already familiar.

The first advantage, in my view, is that a Q-first approach reflects the intellectual structure of contemporary physical chemistry more accurately than a T-first approach. That is, there is an intellectual dynamic in seeing the emergence of the macroscopic from the underworld of atoms. Seeing the properties of bulk matter emerge from the properties of atoms is satisfying, rewarding, and introduces students to the power of scientific explanation. It also enriches the concepts of thermodynamics and illuminates otherwise purely phenomenological concepts. Of course, in the T-first approach it is possible (and in fact desirable) to interpret thermodynamic functions at a molecular level, but with a Q-first approach the links and explanations can be very much deeper.

The second advantage of a Q-first approach is that it opens the door to an early introduction of the modern and enthralling. Thermodynamics is perceived (with let's admit, some truth) to be passé; the modern age is built around quantum theory and its implications for atoms, molecules, and materials. If we want to excite our students, then we are more likely to be able to do so with a quantum than with a thermodynamic function. Through the early introduction of quantum ideas we open the door to the presentation of modern topics, including spectroscopy, molecular reaction dynamics, femtochemistry, computation, and the emerging fields embraced by nanotechnology and nanoscience. In short, we have the opportunity to expose our students to the shock of the new.

The third advantage of Q-first is that it is more e-friendly than thermodynamics. It is, I think, much more open to representation using multimedia, especially computer graphics than thermodynamics is. Of course, graphics and animations can be used in thermodynamics, as in the presentation of three-dimensional surfaces representing equations of state and animations of Carnot cycles and the like, but the real strength of molecular graphics comes from the depiction of wavefunctions, probability densities, and so on, and an additional strength of computation in general comes when it is brought to bear on the Schrödinger equation. I shall discuss the role of graphics in a later section.

Apart from the advantages of a T-first approach that I outlined earlier, it seems to me that there is one serious disadvantage of a Q-first approach, which is the unfamiliarity and depth of the mathematics needed to do anything in quantum mechanics. Great care must be taken in a Q-first approach not to overwhelm students at that early stage of their physical chemistry course (or, indeed, at any stage), especially when heavy mathematics is in alliance with bizarre concepts.

The impact of biology

One of the most important developments in physical chemistry in recent years has been the way in which it has turned its attention to biology. This hugely important step reflects the maturity of the subject: we have cut our teeth on simple systems and are now ready to tackle the complex. There is no doubt that our presentations of physical chemistry should reflect the change in direction of our searchlight's beam, but we should proceed with caution.

The advantages of introducing biological applications are rather obvious and do not need to be rehearsed here. It is enough to say that they are central to modern physical chemistry research and compellingly interesting to most of our students.

They provide a vehicle for demonstrating the vivacity of research in physical chemistry, its relevance, and its challenges. Of course we have to meld biological applications into our presentations.

That, however, is not the only consideration, and we have to be prepared to temper our enthusiasm a little. First, we have to ensure that the presentation of biological applications does not obscure the simplicity of the physical chemistry concepts that we are attempting to convey. The applications of physical chemistry to biology can be rather subtle and demand a familiarity with biological concepts—the components of a cell, the details of biochemical cycles, the structures of biological macromolecules—that may be obscure to our students and suggest to them that physical chemistry is actually harder than it is. Second, there is another wing to physical chemistry: the contribution it can make and is making to materials science, especially the newly emerging aspects of nanotechnology. We should not generate an army of biophysical chemists when we need troops to solve the increasingly important problems associated with new materials.

In short, and as in everything, we need to keep a balance between the temptations of biology and the temptations of materials science. We should certainly show how physical chemistry is relevant to these topics, but not lose sight of the simplicity of the central core of ideas that we are trying to convey. We must educate people into flexibility.

A related issue, which I shall not address here, is the opposite of what I have been addressing. I have been considering the illumination of physical chemistry by biology. The opposite is the illumination of biology by physical chemistry, and the courses and topics that are essential for a biologist (in the broadest sense) to know.

The challenges and the opportunities

Next, I shall examine the challenges that confront us as we try to teach physical chemistry. Figure 1a summarizes what I think are the main difficulties: there is the mathematical aspect of our subject, the abstract character of many of its central concepts, and the overall complexity of physical chemistry. No difficulty is an island, and I like to think that the triangle summarizes the interplay between difficulties rather than their isolation.

We are taught in business school (I am told) that every challenge is an opportunity. That is probably untrue in physical chemistry (and perhaps in commerce too), but there are certainly opportunities for us to enhance our teaching. I have identified three principal ones in Fig.1b, namely graphics, curriculum reform, and the conceptual basis of our subject. As for challenges, no opportunity is an island, and I like to think that the triangle summarizes the interplay between them and the strength that they acquire in combination.

Figure 1c brings the challenges and opportunities together. This illustration will form the structure of my presentation here, for I want to explore how different challenges can be attacked by the various opportunities that we currently have at our disposal.

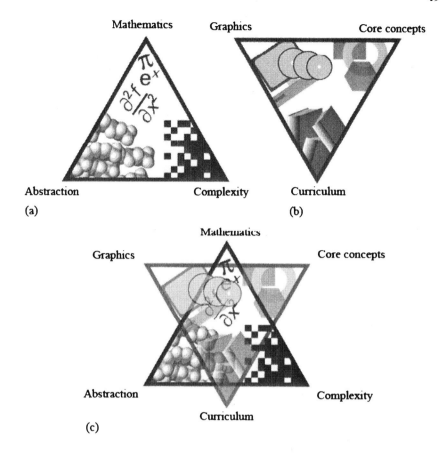

Figure 1. (a) The challenges, (b) the opportunities, and (c) their conjunction.

First, the curriculum. When we are designing our courses we should take into account the techniques that are currently driving our subject. Table 1 summarizes a selection.

Without doubt, to my mind at least, it is computation that is currently having the greatest impact on what we do and how we do it. Computation begins with the control of experiments, the gathering of data, and its manipulation and presentation. Computation also opens up methods of investigation that might be inaccessible experimentally by allowing us to calculate, with increasing reliability, the properties of individual molecules and bulk materials. Computation also allows us to portray properties in a visually assimilable form.

Next in order of importance is magnetic resonance, principally NMR. Apart from its obvious importance for the identification of materials and the information it gives about biological macromolecules in their natural

Table 1. The current experimental motivators

Computation

Nuclear magnetic resonance

X-ray diffraction

Lasers

Electrochemistry

Surface microscopy

Nanotechnology

---

environment and its potential for the realization of quantum computing, NMR is an intellectually interesting technique, for it has grown in complexity over the years, from the primitive flip of a nucleus to the awesomely complicated correlation techniques of today. That intellectual development can be reflected in our teaching. However, NMR does present a real challenge for instructors, for I find it extraordinarily difficult to present pulse techniques and correlation spectroscopy in a visualizable form (presumably because it involves off-diagonal elements of density matrices). I think there is a pressing need to find a compelling technique of visualization for this centrally important technique.

Like group theory, diffraction techniques ebb from and flow into physical chemistry courses. I think X-ray diffraction, at least, should be presented because it is so central to molecular biology and the steady state. Moreover, it provides an excellent opportunity for demonstrating the power of Fourier transforms in the understanding of physical phenomena. Indeed, it could be very interesting to develop a "Fourier" course that embraced diffraction and modern techniques of spectroscopy.

Lasers come next, not because of their intrinsic construction and mode of operation, but because they open up new dimensions of technique, precision, and scale. The experimental technique of physical chemistry that has benefited most from the laser is Raman spectroscopy, which barely existed before their introduction and is now in full flower, showing enormously detailed and interesting information about bulk matter and surfaces. A technique that was essentially invented by the laser is femtochemistry, where we can catch atoms red-handed in the act of reaction. Lasers have brought us right to the heart of reactions, and as such we must build them into our courses.

The classical field of physical chemistry that still has so much to offer is electrochemistry, with its promise of efficient, clean power, the detection and elimination of pollutants, and, through its marriage with nanotechnology, its promise for computation. Surface science, itself intimately involved with electrochemistry as well as catalysis, also barely existed before scanning tunneling techniques opened surfaces to direct inspection and visualization. Suddenly, with STM and its descendants, surfaces have become interesting (they have always been important).

Finally, although nanoscience is perceived by some as a grant-hunting device, that is far from the truth. The investigation of properties unique to nanometer scales is presenting us with fascinating new challenges and—perhaps most important of all—forcing fruitful alliances between different kinds of chemists.

Multimedia and graphics

I want to be even-handed again, but that is hard when considering graphics and its superset, multimedia, for it plays such a compelling role in education, that it is hard to believe that there can be any disadvantages in its deployment. But there are.

The upside of art is compelling. Many chemists think in a visual way, and many for whom English is a second language can assimilate concepts more easily if the words are presented in company with images. I think it most important that we use visual images heavily to convey physical chemistry.

Graphics provide a great opportunity to ameliorate the presentation of mathematics. This it does at three levels. First, and perhaps most importantly, science is currently undergoing a paradigm shift in the manner in which it is done and in terms of what can be explored theoretically. From the analytical derivations that have characterised its procedures since Newton we are increasingly turning to numerical analysis. This has long been the case with computational chemistry, of course, but now we are exploring aspects of complexity and chaos that simply cannot be investigated analytically except in a few remote instances. It is important that we educate our students in these numerical techniques of investigation and portrayal of results. Second, adroit use of graphics can help overcome the abstraction of some mathematical procedures. We should respect that good chemists are often people with strong visual imaginations, and we should build on that by helping them to acquire mathematical skills in as visual a way as possible. Third, the presentation of straight-forward mathematical concepts and manipulations can be made much more interesting by using one of the mathematical software packages that are now available.

I did promise to try to identify disadvantages of multimedia. There are some, although in my view it is the future, and by listing disadvantages I do not wish to give the impression that I am against its implementation: I am for its improvement.

First, it is currently disappointing. The young are currently brought up with amazing graphics in the games they play, and no academic multimedia experience comes remotely close to the quality of the presentation and the graphics. The reasons are largely due to the expense of using a quality graphics engine; but chemists have not yet shown the imagination to take the leap away from the paradigm of the printed page and to develop a stunning yet academically valuable experience. Perhaps the alliance of computer games and intellectual, educational good practice will never be achieved. I doubt that, but I am not aware that it has yet been achieved. Perhaps it must await the current younger generation to come forward with their experience of gaming and apply to problems of education. So, currently, all multimedia experiences are disappointing and add very little to the educational experience.

Second, and for this possibility I have absolutely no evidence: I fear the possibility that multimedia will result in a long-term diminution of imagination. Multimedia might be like watching the movie of the novel rather than reading the novel itself. In the absence of multimedia, students are forced to grapple with concepts, just as the reader of a novel must build pictures of characters and scenes in their mind rather than living off the images devised by the director.

Third, multimedia can corrupt. Because it can generate such impressive images, those images can deflect students from the truth. We have only to think of the compelling imagery of the Bohr planetary atom to appreciate just how deeply a false pictorial model can embed itself in our consciousness. One fear could be that multimedia could spawn a plague of Bohr models.

Fourth, but by no means least, education must be allied to examination, but so far no one has devised a really satisfactory mode of examining material that has been taught by using multimedia. That we educate in a modern mode but examine in a traditional mode is unacceptable, and multimedia education will not flourish until the elusive goal of appropriate examination has been achieved. These remarks apply to components of multimedia, such as the computation of electron densities, surfaces, and dipole moments. How do we (indeed, should we) assess a student's competence at using software packages that are little more than sophisticated black boxes?

Finally, and to some extent trivially and certainly temporarily, multimedia is just plain inconvenient. Until it is implemented in instant-on (currently available to a limited degree) readable in sunlight page-sized screens (coming along) with book-like browsing abilities, it will remain a pain to use informally.

However, it is the future, and once these problems have been addressed and solved, the learning experience will be much richer than it is now. Currently, electronic books are appearing with embedded manipulable graphs and illustrations, with embedded spreadsheets for exploring the consequences of changing parameters, and static illustrations that burst into animation. Multi-level expositions are becoming available, where derivations can be hidden until required, where mathematics background can spring into action as needed, and where the whole reading experience becomes alive.

There is, though, a word of warning that is relevant to the delivery of lectures. The great advantage of chalk is that the lecturer can present an equation as a growing organism and talk through its construction. I think that that approach has more pedagogical impact than simply slapping down a ready made equation on a screen. Of course, multimedia are perfect for exploring the *consequences* and implications of equations, but they can undermine derivations.

On a more positive note, multimedia enable us to respond pedagogically to the shifting paradigms of science, where, as remarked above, numerical analysis is replacing analytical solutions of equations. Through it, we can present computational chemistry, for instance, and pursue the fascinating complexity that arises from systems of coupled differential equations in chemical kinetics.

The central equations

In this final section I shall attempt to identify what I regard as the central equations of physical chemistry, the equations that everyone should remember and appreciate their implications (they are summarized in Table 2). I will identify a handful of equations from each major region of physical chemistry, and finally identify what I regard as the king of all equations.

In what I broadly regard as "equilibrium" (essentially thermodynamics), the two central concepts are the perfect gas equation of state, $pV = nRT$, and the chemical potential, $\mu = \mu^\circ + RT \ln a$. The former is a kind of quide for the formulation of the latter, and the latter, through seeking conditions for the equality of the chemical potentials of substances in various phases underlies the description of all physical and chemical equilibria. These merge in probably the single most useful equation of all chemical thermodynamics, $\Delta_r G^\circ = -RT \ln K$ and the crucially important link between thermodynamics and electrochemistry, $\Delta G = w_e$. The relation that governs the response of systems to changes in the conditions is $dG = Vdp - SdT$, and summarizes almost everything we need to know.

In what I broadly regard as "structure" (essentially quantum theory), the equation that epitomizes the transition from classical mechanics to quantum mechanics, is the de Broglie relation, $\lambda = h/p$, for it summarizes the central concept of duality. Stemming from duality is the aspect of reality that distinguishes quantum mechanics from classical mechanics, namely superposition: $\psi = \psi_A + \psi_B$ with its implication of the roles of constructive and destructive interference. Then of course, there is the means of calculating wavefunctions, the Schrödinger equation. For simplicity I will write down its time-independent form, $H\psi = E\psi$, but it is just as important for a physical chemist to be familiar with its time-dependent form and its ramifications for spectroscopy and reaction.

Table 2. The core equations

Thermodynamics	$pV = nRT$
	$\mu = \mu^\circ + RT \ln a$
	$dG = Vdp - SdT$
	$\Delta_r G^\circ = -RT \ln K$
	$\Delta G = w_e$
Structure	$\lambda = h/p$
	$\psi = \psi_1 + \psi_2$
	$H\Psi = i\hbar \partial \Psi / \partial t$
Change	$v = f([A], [B],...)$
	$k = A \exp(-E_a/RT)$
Overall	$p_i = \exp(-\varepsilon_i/kT)/q$

In what I regard as the world of "change" (essentially chemical kinetics and dynamics), there are three central equations. One is the form of a rate law, $v = f([A],[B]...)$, and all its implications for the prediction of the outcome of reactions, their mechanisms, and, increasingly, nonlinear phenomena, and the other closely related, augmenting expression, is the Arrhenius relation, $k = A\exp(-E_a/RT)$, and its implications for the temperature-dependence of reaction rates. Lurking behind discussions of this kind is the diffusion equation, in its various flavors starting from the vanilla $\partial P/\partial t = -\partial^2 P/\partial t^2$ (which elsewhere I have referred to as summarizing the fact that "Nature abhors a wrinkle").

I promised to end with the king of equations, the last to be thrown from our balloon. In my view, this king must be the Boltzmann distribution, $p_i = \exp(-\varepsilon_i/kT)/q$. To a large extent, this equation summarizes the whole of chemistry. First, it is derived from the most primitive of assumptions (the random distribution of entities over available energy levels subject to a couple of obvious constraints) and consequently applies universally. Because it uses the concept of energy levels yet can be used to calculate thermodynamic properties, it is a bridge between the microscopic and macroscopic worlds. Third, because it is essentially a way of distinguishing populations that are stable ($\varepsilon_i < kT$) from those that are volatile ($\varepsilon_i > kT$), it captures the essence of the two pillars of chemistry: structure and change.

Conclusions

Before I summarize these remarks, I would like to make one further plea. As educators of huge numbers of chemists (and in particular physical chemists), the US educational system has a responsibility to propagate good practice. "Good practice" does not mean what one was taught oneself decades ago; it means what has been agreed for notation, nomenclature, and symbolism by the international community through IUPAC, to which the USA is a major and influential member. A major generator of the chemical literature, from textbooks up to primary research papers, has a duty to communicate in the accepted language of the international community. So, whatever the strategic decisions taken about the future of education in physical chemistry, instructors and authors have a duty to express themselves in the international language of science. It seems to me, for instance, to provide a tiny example, absurd that the most modern branch of physical chemistry (computational chemistry) in large measure still reports heats *(sic)* of formation in kilocalories per mole. There are numerous examples throughout the literature of the subject that make use of obsolete units and incorrect modes (that is, not in conformity with agreed international practice) of manipulation and presentation. Many *think* they know correct usage; but the evidence is that they do not. Adherence to IUPAC recommendations greatly simplifies the assimilation of our subject by our students, makes it easier for our students to move between topics of chemistry and branches of science, and propagates the responsible thought that they are members of a universal community.

The centrally important point of these reflections, however, is the strategy of education in physical chemistry. We should identify the core ideas that we wish students to carry away from our courses, and identify the current motivators of our subject. We should use multimedia wherever appropriate (but not to excess) and develop ways of examining the new skills that multimedia brings in its train. We should be sensitive to the difficulty that large numbers of students have with mathematics, and never fail to interpret the salient features of the equations we derive. Graphics is perhaps the savior of chemistry, for its adroit use can be used to generate insight (or, at least, to transmit our insights) and to find ways to assimilate mathematics.

Our examples should be drawn from the frontiers of physical chemistry, to show that it is still a dynamic, living subject even though its underlying principles are now a century old. Those examples should include biological applications, but not to the exclusion of the contributions that physical chemistry makes to other branches of science and daily life.

Philosophy of Physical Chemistry

Chapter 6

Philosophy of Chemistry, Reduction, Emergence, and Chemical Education

Eric Scerri

Department of Chemistry and Biochemistry, University of California at Los Angeles, Los Angeles, CA 90095 (scerri@chem.ucla.edu)

Introduction

In this article I hope to introduce chemical educators to some of the work carried out in the philosophy of chemistry. The relevance of such work and especially that carried out on the reduction of chemistry to physics is considerable, and especially so in the case of physical chemistry. As the very name of the discipline implies, physical chemistry juxtaposes aspects of chemistry with aspects of physics. The relationship between these two classical areas of science needs to be considered in order to ascertain the extent to which chemistry should be taught as applied physics or to inform the teaching of physical chemistry per se.

The customary beginning of discussions on the reduction of any specific scientific field is to assume a form of hierarchy among physics, chemistry and biology. Biology deals with complex living systems, whose inner workings can be studied by looking to chemistry. It has become something of a truism these days that the discovery of the structure of DNA in the early 1950s has been perhaps the single most important scientific discovery of the twentieth century. The subsequent strides achieved in molecular biology would seem to indicate that biology straightforwardly reduces to chemistry. But of course not everything about living systems is necessarily a matter of chemistry. The most global aspect that all living systems possess is just the fact that they are alive. To many commentators this attribute is not just the sum of so many chemical parts working together. Poets, mystics and theologians tell us that life is sacred. Philosophers, or at least some of them, tell us that consciousness, a property found in many living systems, is not reducible to chemical or biochemical goings on.

Nevertheless, the study of the relationship between chemistry and physics has also provided a huge boost to the advocates of reduction. For example, the development of quantum mechanics experienced some of its earliest successes in providing an explanation for chemical phenomena. To this day among the biggest users of quantum mechanical methods are computational and/or theoretical chemists. While Niels Bohr began to establish the electronic configurations of atoms, the explanation of the periodic system was taken a great deal further by the work of another quantum physicist, Wolfgang Pauli. A little later the chemical bond, which was regarded merely as a mysterious pairing of electrons by G.N. Lewis, was given a fundamental underpinning by quantum mechanics. Bonding could now be regarded as resulting from the simultaneous constructive and destructive interference of waves of electrons emanating from adjacent atoms.[i] Why then should anyone in their right mind question the notion that biology reduces to chemistry and that chemistry reduces to physics? One response to this question is that attempts at reduction usually lead to progress but are seldom as complete as their proponents would have us believe they are.

The explanation of the periodic system by quantum mechanics, for example, is only partial. The possible lengths of the various periods in the table follow deductively from the solution of the Schrödinger equation for the hydrogen atom and the relationship between the four quantum numbers, which is also obtained deductively. However, the repetition of all but the first period length remains a source of debate (1). The repetition of all the other period lengths has not been deduced from first principles however (2). Stated more precisely, the empirical order in which the atomic orbitals are filled has not been deduced. If this were possible the explanation for the lengths of successive periods, including the repetitions, would follow trivially.

But to return to the general question of the reduction of chemistry to physics, this is a topic that is of practical as well as intellectual interest. It is important for educators to reflect upon and moreover to begin to address explicitly when it comes to curricular development and course design. Although the present article will not be addressing the latter issue explicitly, it is gratifying to report that some authors have begun the task of introducing the notion of the reduction of chemistry explicitly into chemistry courses (3, 4, 5).

How philosophers of science have approached the reduction of chemistry to physics

The establishment of the reduction of one scientific discipline to another one by philosophers of science requires more than an appeal to the importance of DNA in biology or success of quantum chemistry. It requires that formal relations be established between the two fields in question. At least this has been the requirement imposed by the Logical Positivist school of philosophers which was responsible for the professionalization of philosophy of science.[ii] According to this view, the philosopher is required to axiomatize a chemical theory as well as the putative reducing theory from physics. She is then required to establish some 'bridge principles' which allow one to connect the two levels of discourse in a formal logical manner. However, although much work has been expended on this form of reduction, the consensus view is that there is not a single case in which such rigorous connections have been established.

Even the case of reducing classical thermodynamics to statistical mechanics, which is often touted as a paradigmatic example of strict reduction, has been found to be somewhat problematic (6). Following the general trend away from formal logic and towards 'naturalism' that has occurred in philosophy of science, it has been proposed that the reduction of chemistry should be examined in the same manner that a chemist or physicist might (7). This involves considering the extent to which a branch of chemistry, say chemical kinetics, can be deduced from the appropriate underlying physical theory or quantum mechanics in this particular case. The outcome of following such an approach has been the conclusion that there exists an approximate reduction of chemistry to quantum mechanics, augmented by special relativity in some cases. Not surprisingly the reduction is by no means exact or complete.

Here is one immediately relevant point for chemical educators. These studies carried out in philosophy of chemistry remind us of the approximate nature of physical accounts of chemical phenomena and they document the 'reductive gap' that remains to be bridged, something that quantum chemists and other practitioners are not generally too eager to discuss in publications.

Inorganic chemistry, for example, is increasingly dominated by approximate models from quantum mechanics. If one were to open any chemistry textbook to the chapter on transition metals, one would find extensive discussions of crystal field theory and details of how the 5-fold d orbital degeneracy is removed by ligands depending on whether their geometrical arrangement is octahedral, square planar, tetrahedral or linear. The actual chemistry of the elements is neglected at the expense of 'principles' which almost invariably means approximate models. The excitement of chemistry tends to be devalued at the expense of formal rules.[iii] In spite of many articles from chemical educators who bemoan this situation, the physics-driven approach retains its stranglehold over chemistry and few textbooks dare to break away from this approach.

Epistemological and Ontological Reduction

Let us now delve a little deeper into the philosophy of chemistry. As mentioned above, the conclusion of those working in the field has been that chemistry does not fully reduce to quantum mechanics but that this is a matter of epistemology and not of ontology. In other words, while it is accepted that theories of chemistry do not reduce to theories of physics many philosophers of chemistry have specifically stated that they believe that chemical entities are nothing but physical entities at bottom. After all, the failure to establish epistemological reduction may be due to some limitations in our current understanding of chemistry and physics or even both of them. Surely, or so the previous thinking has been, chemistry is nothing more than physics from a 'God's eye' point of view, or in other words from an ontological perspective.

A more general approach adopted by many philosophers is called physicalism. This does not entail reduction of the theoretical or epistemological kind but is a question of whether the physical determines the chemical. Physicalism has been invoked in philosophy of mind to argue that there is a dependence relationship between mental events and physical goings on in the brain. This is clearly a weaker form of reductive claim that epistemological reduction discussed earlier in this article.

But physicalism has come under severe criticism and is now seen as suffering from two major drawbacks. In order to defend the claim that mental or chemical properties are determined by the physical, when this cannot yet be achieved by recourse to current physics, the defendant of physicalism appeals to a future perfected physics. However this claim appears to be vacuous because we may never reach the point of a perfected physics.

The second objection is that a mere dependence of the chemical say on the physical can work both ways. The chemical may determine the physical or vice versa. To have any validity the physicalist needs to argue that the dependence is asymmetrical and only functions in the sense that the chemical is dependent upon the physical and not vice versa.

Various responses

Quite recently this matter has come to the surface in the course of an article in which authors Lombardi and LaBarca explicitly point out that most previous workers in the philosophy of chemistry have assumed that ontological reduction necessarily holds regardless of the fortunes of epistemological reduction (8). Whereas the distinction between epistemological and ontological reduction is by no means universally agreed upon, perhaps a few general words might help to orient the general reader. Broadly speaking epistemological reduction has been taken to refer to the question of whether theories of the allegedly reduced science such as chemistry can be reduced to the theories of the reducing science, namely physics. Ontological reduction is not exhausted by our knowledge of any two particular sciences. It is concerned more with the question of whether the entities of the allegedly reduced science are reduced to the entities of the reducing science. Different authors differ regarding the extent to which they are prepared to admit theoretical knowledge enter into such discussions.

Lombardi and LaBarca have pointed out any adherence to the view that chemistry is ontologically reduced to physics, regardless of the outcome of epistemological reduction is nothing but an article of faith. As they further ask, "what if even epistemological reduction of chemistry to physics does not in fact hold good?" More importantly perhaps, their raising this question has led to a discussion of how one is to assess the success or otherwise of the ontological reduction of chemistry to physics. How does one actually do ontology?

Responses to this further question appear to fall into two camps. One prominent metaphysician believes that the question needs to be approached independently of any theories of chemistry and of physics. Robin Le Poidevin has published an extensive article in which he argues in favor of the ontological reduction of chemistry to physics. He does this through what he has termed a combinatorial approach.

Le Poidevin directs his approach at the periodic system of the elements. He writes,

> Why was Mendeleev so confident that the elements he predicted actually existed? This is not a question about his confidence in their periodic law (that, as he formulated it, 'the elements, if arranged according to their atomic weights, exhibit an evident periodicity of properties' (Mendeleev [1889], p. 635)), but rather about an implicit conceptual move. Granted that the gaps in Mendeleev's table represented genuine possibilities, elements that could exist, why assume that the possibilities would [be][iv] realized? (9).

> Another question is related, but purely philosophical: even if some elements in the table are merely possible, there is a genuine difference between the physical possibility of an element between, say, zinc and arsenic (atomic numbers 30 and 33), and the mere logical possibility of an element between potassium and calcium (19 and 20) (10).

Le Poidevin is fully aware of the problem whereby dependence can function both ways. The mere physicalist hope that the physical determines the chemical rather than vice versa will not convince anyone who doubts this point. Le Poidevin wants to put these doubts to rest by appeal to what he calls 'combinatorialism'.

> The central contention of combinatorialism is this: possibilities are just combinations of actually existing simple items (individuals, properties, relations). Let us call this the principle of recombination. To illustrate it,

suppose the actual world to contain just two individuals, a and b, and two monadic properties, F and G, such that (Fa & Gb). Assuming F and G to be incompatible properties, and ignoring the possibility of there being nothing at all, then the following is an exhaustive list of the other possibilities:

Fa
Fb
Ga
Gb
Fa & Fb
Ga & Gb
Ga & Fb

1-6 are 'contracted worlds', ones that lack one or more components of the actual world (*11*).

Le Poidevin explains that combinatorialism is a form of reductionism about possibilia. He claims that the talk of non-existent possibilia is made true by virtue of actual objects and their properties, just as the inhabitants of model world mentioned above is made possible by virtue of a and b and the properties F and G. Presumably the reader is being invited to also consider such examples as Mendeleev's predicted elements in this way. According to le Poidevin's approach, the elements that are as yet non-existent but physically possible are those that can be regarded as combinations of some undefined basic objects and/or basic properties.

Le Poidevin suggests that this approach provides a means of establishing the required asymmetry in order to ground the reduction of the chemical to the physical or the mental to the physical, and a means of countering the symmetry problem alluded to earlier.

...granted that some property F, can be identified with some physical property G, what makes it the case that G is a more fundamental property than F? that F is determined by G, and not vice versa? Here is the combinatorial criterion:

A property-type F is ontologically reducible to a more fundamental property-type G is the possibility of something's being F is constituted by a recombination of actual instances of G, but the possibility of something's being G is not constituted by a recombination of actual instances of F (*12*).

Although Le Poidevin is trying to establish the ontological dependence of the chemical upon the physical, he is attempting to do this without appeal to chemical or physical theories. By avoiding the use of physical theory he hopes to avoid the usual pitfalls of physicalism. Moreover he wishes to distinguish clearly between ontology and metaphysics and what better way to achieve this than by avoiding the results of scientific theory or indeed an appeal to any scientific theories whatsoever.

Le Poidevin writes that because the thesis of ontological reduction is about properties, we do have to have a clear conception of what is to count as a chemical property. He then takes the identity of an element, as defined by its position in a periodic ordering, and its associated macroscopic properties to be paradigmatically chemical properties.   About these properties we can be unapologetic realists. He also claims that a periodic ordering is a classification rather than a theory, so this conception of chemical properties is as theory-neutral as it can be.[v] He believes that the question of the ontological reduction of chemistry is the question of whether these paradigmatically chemical properties reduce to more fundamental properties. He then adds,

> What counts as a fundamental property? One criterion is the combinatorial one of the previous section: the more fundamental properties are those properties combinations of which generate the range of physical possibilities of the less fundamental properties (*13*).

> We might, just accept it as a brute fact about the world that the series of elements was discrete. But if there were a finite number of properties, combinations of which generate the physical possibilities represented by the periodic table, then variation would necessarily be discrete rather than continuous. We can believe in the existence of these fundamental entities and properties without subscribing to any particular account of them (e.g. an account in terms of electronic configuration), such accounts at least show us the way in which chemical properties could be determined by more fundamental ones. The point is that, given the principle of recombination, unless those more fundamental properties exist, unactualized elements would not be physical possibilities (*14*).

> Let me try to rephrase the argument. We assume that the combination of a finite number of fundamental properties, via a combinatorial approach, leads to a discrete set of macroscopic physical possibilities. We also know empirically that the chemical elements occur in a discrete manner because there are no intermediate elements between, say, hydrogen and helium. The combinatorial approach can thus be taken as an explanation for the discreteness in the occurrence of elements and furthermore it justifies the fact that Mendeleev regarded the yet undiscovered elements like germanium as being physical possibilities rather than merely logical ones.

To give Le Poidevin due credit, I think he has provided a plausible argument for the ontological reduction of the existence of the chemical elements and he has done so without any appeal to physical theory, either present or future. And in doing so he has shown us an interesting way in which questions of ontological reduction can be treated independently of the theories of chemistry or physics. But as to whether he has separated the question of ontological reduction as fully from that of epistemological reduction as he seemed to promise in his article, this is debatable. Admittedly, the ordering of the chemical elements may not be in any sense theoretical, as he states, but there is no denying that ordering the elements by way of atomic number, or by whatever other means, is dependent on our knowledge of the elements. It is just that this knowledge takes the form of a classification or ordering rather than a theory as Le Poidevin correctly points out. But surely this does not render the act of classification any less epistemological.

Finally, I would like to point out some specific points concerning Le Poidevin's analysis. Let me return to the question of the discrete manner in which the elements occur. This fact, it will be recalled, is taken by Le Poidevin to support a combinatorial argument whereby a finite number of fundamental entities or properties combine together to give a discrete set of composite elements. But what if we consider the combination of quarks (charge $= 1/3$), instead of protons (charge $= 1$)? In the former case a finite number of quarks would also produce a discrete set of atoms of the elements, only the discreteness would involve increments of one-third instead of integral units.

And if this were the case, then it would be physically possible for there to be two elements between say $Z = 19$ and $Z = 20$ to use Le Poidevin's example.[vi] Let us further suppose that a future theory might hold that the fundamental particles are some form of sub-quarks with a charge of 0.1 units. Under these conditions combinatorialism would lead to the existence of nine physical possibilities between elements 19 and 20, and so on. It would appear that Le Poidevin's distinction between a physical possibility, as opposed to a merely logical one, is dependent on the state of knowledge of fundamental particles at any particular epoch in the history of science which is surely not what Le Poidevin intends. Indeed the distinction proposed by Le Poidevin would appear to be susceptible to a form of vacuity, not altogether unlike the threat of vacuity which is faced by physicalism, and which was supposed to be circumvented by appeal to combinatorialism.

Let us return to Le Poidevin's sense of physical possibility once again. As the reader will recall, Le Poidevin claims that there is no physical possibility of an element between $Z = 19$ and $Z = 20$ whereas Mendeleev would have been confident of physical possibilities between elements $Z = 30$ and $Z = 33$. However, it is well known to historians of science that Mendeleev was an ardent opponent of any form of atomic sub-structure. Mendeleev himself would not therefore have availed himself of a combinatorial argument in order to propose the existence of one or two physically possible elements in this gap in the sequence of the elements.[vii] Mendeleev's argument was altogether much simpler. He noticed a considerable gap between the atomic weights of zinc and arsenic (Z = 30 and 33 respectively) and together with the aid of some vertical trends in the surrounding elements he was able to predict rather accurately the properties of two missing elements.

Finally there is a somewhat general objection to the use of combinatorialism in order to ground the ontological reduction of chemistry. Surely the assumption of that fundamental entities combine together to form macroscopic chemical entities ensures from the start that the hoped for asymmetry is present. But it seems to do so in a circular manner. If one assumes that macroscopic chemical entities like elements are comprised of sub-atomic particles then of course it follows that the reverse is not true. The hoped for asymmetry appears to have been written directly into the account, or so it would seem.

Physicalism and McLaughlin's analysis of Broad's Emergentism

Meanwhile several other authors in the field believe that one may consult the findings from scientific theories in order to discuss ontological questions and indeed that our only access to ontology is through the findings from scientific theories. They do not believe that there is any such thing as pure ontology, which can be examined without recourse to scientific findings.[viii] Rather they hold that to examine the ontology of chemistry and physics requires an understanding of what contemporary scientific theories tell us about the nature of the fundamental entities in chemistry and physics respectively.

Let us take up again the question of symmetry which was discussed earlier in the present article. The notion that the physical may be dependent upon the chemical may seem unusual to those who are unfamiliar with these philosophical arguments. It is after all taken as a foregone conclusion in physics and chemistry that the dependence works in one direction only, namely that matters at the physical level determine what happens at the chemical level and not vice versa. But it is precisely this question, which is being challenged when we begin to ask the philosophical questions about reduction and dependence of levels.

The worry about whether the dependence relationship may be symmetrical or whether it is driven from top to bottom has come mainly from philosophers working in the area of philosophy of mind. Here is one example, which has often been discussed. Let us suppose I am working on an article for publication and that I decide that I need to visit the research library at the other end of campus. If I follow this intuition a whole number of subsequent steps are put into motion. I need to lock my office door, pick up my sunglasses and begin to make my way across to the library. What is the dependence relationship between the different levels of organization in my body following decisions of this kind? It would not seem implausible to say that my decision to go to the library, a mental event, has led to a number of physiological movements, my walking across campus, which in turn has led to much chemical and neurological activity within my body and ultimately to a change in the manner in which the physical components making up my body are obliged to move if I am to make it to the other side of campus. It is in this general sense, motivated by questions in the philosophy of mind that a reversal of the usual chain of dependence of levels appears to be rather uncontroversial.

Would anyone have doubted the usual direction of dependence of levels by considering a chemical reaction? For example when compound A and B react are they reacting because they are being driven to do so from below by the protons and electrons in their molecules? Or are they reacting because of some property that is possessed by the two molecules at the chemical level of organization? What is the direction of dependence? It is by no means clear that reactivity is driven from below, that the chemical is dependent on the physical and not vice versa. If indeed the chemical levels determine the physical it is said that 'downward causation' is taking place.

The philosopher Brian McLaughlin has examined the views of emergentists, more specifically what he terms the British Emergentists of whom C.D. Broad was a leading example. For emergentists like Broad downward causation is possible in addition to the more usual upward causation. In addition to physical levels determining what occurs at the chemical level, the emergentist like Broad claims that there also exists downward causation as evinced by certain phenomena like chemical bonding. Indeed Broad speaks of there being 'configurational forces' which result from a particular arrangement of particles over and above resultant forces due to the pair-wise interactions which can be accounted for in terms of the fundamental forces of nature such as gravitational and electromagnetic forces.

Broad believed that emergent and mechanistic chemistry (non-emergent chemistry) agree in the following respect,

> That all the different chemical elements are composed of positive and negative electrified particles in different numbers and arrangements; and that these differences of number and arrangement are the only ultimate difference between (15).

Nevertheless he wrote that if a mechanistic (non-emergent) chemistry were true,

> ...it would be theoretically possible to deduce the characteristic behavior of any element from an adequate knowledge of the number and arrangement of

the particles in its atom, without needing to observe a sample of that substance. We could, in theory, deduce what other elements it would combine with and in what proportions; which of these compounds would react in the presence each other under given conditions of temperature, pressure etc. And all this should be theoretically possible without needing to observe samples of these compounds (*16*).

The situation with which we are faced in chemistry…seems to offer the most plausible example of emergent behaviour (*17*).

But McLaughlin, writing in 1992 denies the presence of any configurational forces suggested by Broad and denies the possibility of emergence by referring to the success of the quantum mechanical theory of bonding which had not yet taken place when Broad was writing.

It is, I contend, no coincidence that the last major work in the British Emergentist tradition coincided with the advent of quantum mechanics. Quantum mechanics and the various scientific advances made possible are arguably what led to British Emergentism's downfall…quantum mechanical explanations of chemical bonding in terms of electromagneticism [sic], and various advances this made possible in molecular biology and genetics – for example the discovery of the structure of DNA – make the main doctrines of British emergentism, so far as the chemical and the biological are concerned at least, seem enormously implausible. Given the advent of quantum mechanics and these other scientific theories, there seems not a scintilla of evidence that there are emergent causal powers or laws in the sense in question… and there seems not a scintilla of evidence that there is downward causation from the psychological, biological and chemical levels (*18*).

McLaughlin takes the fact that there is now a highly successful quantum mechanical account of chemical bonding to indicate that the chemical level is dependent upon the physical and that physical forces bring about bonding at the chemical level. As may be seen in the above quotation McLaughlin takes this state of affairs to leave no room whatsoever for downward causation.

What he does not seem to realize is that a perfectly good explanation existed for chemical bonding prior to the advent of the quantum mechanical explanation, namely Lewis's theory whereby pairs of electrons form the bonds between the various atoms in a covalently bonded molecule. Although the quantum mechanical theory provides a more fundamental explanation in terms of exchange energy and so on is undeniable but it also retains the notion of pairs of electrons even if this notion is now augmented by the view that electrons have anti-parallel spins within such pairs.

Moreover McLaughlin seems to assume that the explanation for bonding provided by the quantum mechanical theory is complete. Although it certainly is an advance on Lewis's theory Broad's point that one cannot predict the properties of a compound from the properties of the combining elements is not completely addressed in the newer theory either. Although one may use quantum mechanics to predict such properties as bond lengths or dipole moments in a molecule that is known to form one cannot predict which particular molecules will form in general, although one may obtain estimates of the relative stabilities of the likely products. McLaughlin is clearly not aware of the current status of calculations in quantum chemistry and is making an over-general statement.[ix]

Conclusions

The reduction of chemistry to physics or more specifically to quantum mechanics is an important topic for chemical educators, particularly those concerned with physical chemistry and general chemistry. This article has sought to discuss various facets of the question of reduction including some recent work by philosophers on the question of the ontological reduction of chemistry and the question of whether chemistry emerges from physics in any sense. The two authors cited in this context, Le Poidevin and McLaughlin have each concluded that chemistry is reduced to physics for rather different reasons. McLaughlin has concluded that the development of a quantum theory of chemical bonding also renders the notion of the emergence of chemistry as envisaged by authors such as C.D. Broad completely implausible. The present author believes that these two philosophers are mistaken in drawing these conclusions, partly because they are not sufficiently familiar with the scientific developments that they are discussing.

It is my belief that chemistry does not fully reduce to physics in an epistemological sense and perhaps even fails to reduce in an ontological sense although I have not provided any positive arguments in favor of the latter claim here. I have merely provided a critique of two leading authors who claim to provide conclusive arguments in favor of the ontological reduction of chemistry and the impossibility of emergence. I have also tried to stimulate greater interest among the chemical education community in contemporary issues in the philosophy of chemistry. Given the considerable developments that have taken place within the philosophy of chemistry, especially over the question of the reduction of chemistry, or at least the extent of reduction, it is high time for chemical educators to begin to incorporate some of the conclusions into chemistry courses and chemistry textbooks. Such an addition would surely serve chemical education more effectively than the obligatory, and frequently outdated, accounts of the scientific method which appear in the opening pages of nearly all chemistry textbooks.

Notes

i   I am referring to the molecular orbital approach and not valence bond theory.

ii   Historically the question of reduction goes a good deal further back. Comte's earlier positivism already had an explicitly reductionist agenda. I am grateful to a reviewer for his input on this point.

iii   Somewhat curiously, the task of dwelling on the 'real chemistry' of the elements for example is left to amateurs such as Oliver Sacks whose book *Uncle Tungsten* has been praised by numerous professional chemists (*19*)

iv   Le Poidevin actually wrote "were realized". I believe he means to say, "would be realized".

v   It may well be theory neutral but it is surely not epistemologically neutral. God does not order the elements. It is scientific knowledge by way of Moseley and others that has allowed us to order the elements into a coherent sequence. Again I don't want to deflect any attention from the main line of argumentation.

vi   The possible existence of quark matter, meaning elements with fractional atomic charges has been examined in a number of publications. Among them are some articles by C.K. Jorgensen (*20*).

vii   Of course from a strictly normative point of view this point will not carry much weight.

viii   This is in keeping with the kind of philosophical naturalism that has been championed by leading philosopher Quine.

ix   McLaughlin is also mistaken in believing that the discovery of the structure of DNA owes anything whatsoever to the quantum mechanical theory of bonding. The only possible connection might be that Linus Pauling was involved in both developments although by his own admission he failed to discover the structure of DNA. However in his attempts as well as in the earlier discovery of the structure of protein molecules Pauling did not draw in any way on quantum mechanics.

References

1. Scerri, E.R. Foundations of Chemistry, **2004**, *6*, 93-116.
2. Scerri, E.R. In *The Periodic Table: Into the 21st Century*, Editors, Rouvray, D. ; B. King.; Science Research Press, UK, 2004; pp 142-160.
3. Erduran, S. *Science Education*, **2001**, *10*, 581-594.
4. Erduran, S.; Scerri, E.R. In *Chemical Education: Towards Research-Based Practice*, Editors Gilbert, J.; De Jong, O.; Justi, R.; Teagust, D.F.; Van Driel, J.H. Kluwer, Dordrecht, 2002, pp 7-27.
5. Gimbel, S.; Wedlock, M. *J. Chem. Educ.*, **2006**, *83*, 880-882.
6. Needham, P. In For Good Measure: Philosophical Essays Dedicated to Jan Odelstad on the Occasion of His Fiftieth Birthday, Editors Lindahl, L.; Needham, P.; Sliwinski, R. Philosophical Studies, vol. 46, Uppsala.
7. Scerri, E.R. In *PSA 1994 vol 1*, Editors Hull, D.; Forbes, M.; and Burian, R. Philosophy of Science Association, East Lansing, MI, 1994, pp 160-170.
8. Lombardi, O, M.; Labarca, M. *Foundations of Chemistry*, **2005**, *7*, 125-148.
9. Le Poidevin, R. *British Journal for the Philosophy of Science*, **2005**, *56*, 117-134, *p. 118.*
10. Reference 9, p. 119.
11. Reference 9, p. 124.
12. Reference 9, p. 129.
13. Reference 9, p. 131.
14. Reference 9, pp. 131-132.
15. C.D. Broad. *The Mind and Its Place in Nature*, Keegan Paul, Trench and Trubner, London, 1925, p. 69.
16. Reference 15, p. 70.
17. Reference 15, p. 65.
18. McLaughlin, B. 1992, The Rise and Fall of British Emergentism, In *Emergence or Reduction? Essays on the Prospect of a Non-Reductive Physicalism*, Beckerman, A.; Flohr, H.; Kim J., Eds.; Walter de Gruyter, Berlin, 1992, pp. 49-93, quoted from p. 54-55.
19. Sacks, O. *Uncle Tungsten*, Alfred Kopf, New York, 2001.
20. Jorgensen, C.K. *Structure and Bonding*, **1987**, *34*, 19-38.

Teaching Literature Reviews

Chapter 7

Teaching and Learning Physical Chemistry: A Review of Educational Research

Georgios Tsaparlis

Department of Chemistry, University of Ioannina, Ioannina, Greece

Science education research focuses on studying variables relating to science content in connection with the process of learning. In this chapter, we examine the role of science education theories and tools for the teaching and learning of physical chemistry. We review research work on physical chemistry concepts, with emphasis on the areas of thermodynamics, electrochemistry, and quantum chemistry. Advances in problem solving research, and problems of the conventional expository physical chemistry laboratory are discussed. Finally, context-based approaches to teaching, the role of new educational technology, and active and cooperative learning are covered.

Introduction

Research into science and chemistry education aims at advancing pedagogical knowledge through experimental educational investigation. Chemistry education is closely related to the science of chemistry, and so is research in the two fields (*1*). There is, however, a fundamental difference: chemistry education research focuses on understanding and improving chemistry learning by studying variables relating to chemistry content or to what the teacher or student does in a learning environment (2). It involves "a complex interplay between the more global perspective of the social sciences (i.e., the process of learning) and the analytical perspective of the physical sciences (i.e., the content)."

The *Task Force on Chemical Education Research* of the American Chemical Society has defined the elements of scholarship in chemistry education (3). The following areas were considered: scholarship of teaching; scholarship of discovery; scholarship of application. Characteristics of research are that it is theory based, data based, and produces generalizable results. Of central importance is the support of research with suitable theory or theories, otherwise it would not be different from journalism (4). Johnstone (5) argued that research has provided us with the tools "to harmonize a logical approach to our subject with a psychological approach to the teaching of our subject so that young people will catch our enthusiasm and enjoy the intellectual stimulus which our subject can, and should, offer."

Theories in Science Education

Traditional educational theory and educational practice favored (and practice still does today) the direct transmission of intact knowledge from the mind of the knowledgeable (the teacher) to the mind of the ignorant (the student). Accordingly, knowledge should be judged in terms of whether it is true or false (6).

The advent of Piagetian cognitive psychology caused a paradigm shift (in the Kuhnian sense) in the research programs of science education. Central to Piaget's description of cognitive development was the active construction of knowledge by the individual, a construction that is taking place through a dynamic process of interaction with the world (7-9). To bring about development (*accommodation*) of new ideas (new *schemas*), it is necessary to bring about *disequilibrium*.

Piagetian theory "has had a profound effect on the way we think about learners and learning as well as the methods by which this dynamic development process is translated to learning environments" (10). It dominated science education research for many years, becoming internal to science education; thus, not only students were categorized as concrete or formal thinkers, but also chemical concepts were distinguished into concrete and formal (8, 11). In the opinion of this author and others (10, 12) it still has important messages to convey to science and chemistry educators.

During the past twenty-five years researchers studied concepts systematically. It is now widely accepted that students' scientific concepts are often at odds with the accepted scientific views. A coherent research movement

was formed, the *alternative frameworks* or *alternative conceptions* or *students'
misconceptions* movement (*13, 14*). This movement needed a theory to back it,
and found a suitable and good one in *constructivism*. According to it, knowledge
is constructed in the mind of the learner. It is seldom (or even never) transferred
intact from the mind of the teacher to the mind of the student. It "results from a
more or less continual process in which it is both built and continually tested.
Our knowledge must be viable; it must work ... (and) function satisfactorily in
the context in which it arises"(*6*).

Constructivism is linked to the philosophical-epistemological theory of
(scientific) relativism or empiricism, which is in contrast to another theory, that
of *(scientific) realism* or *objectivism* or *positivism*. Realists believe "logical
analysis applied to objective observations can be used to discover the truth about
the world we live in. Relativists accept the existence of a real world, but question
whether this world is 'knowable'. They note that observations, and the choice of
observations to be made, are influenced by the beliefs, theories, hypotheses, and
background of the individual who makes them" (*6*). Realism and empiricism
must be considered as two extremes on a continuum. It is certain that in its early
years (surely until, say, the beginning of the twentieth century), science was
closer to relativism, but as time passed, we came closer to a *realist* state. Note
that are there are those who argue that philosophical and educational
constructivisms are intertwined (*15, 16*), and those who are against (*17*) and
critical (*18*) of the way these philosophical theories are linked to education
theories.

Educational constructivism extended the realism-empiricism dichotomy into
how individuals learn, and assumed two main forms (*19*): (i) *personal
constructivism*, which is associated with Piaget (*8*); and (ii) *social-cultural
constructivism* which is linked to Vygotsky. Piagetian constructivism is
associated with an idealized person (the 'epistemic subject'). On the other hand,
according to Vygotsky (*20*), the learner constructs actively his/her knowledge,
but this process is greatly assisted by interactions with peers and with the teacher
who acts at the students' *zone of proximal development*.

There are further variants of constructivism (*6*). Ernest von Glasersfeld's
radical constructivism (*21, 22*) extended Piagetian constructivism by assuming
that there is also the person who "does the actual constructing of an epistemic
subject *qua* abstraction, and this is always a psychological subject with all its ties
to the social and historical context in which it operates" (*16*). Kelly's theory of
personal constructs (*23*) has led to an alternative form of personal
constructivism (*6*). Its main hypotheses are first that individuals differ from each
other in their construction of events, and second that one individual's constructs
are similar to another's. The latter hypothesis gives importance to social
interaction, combining personal and social constructivism. For radical and some
social constructivists (such as Kolb - see Discussion), "experience is the ultimate
arbiter for deciding between scientific theories and how students acquire

knowledge" (*16*). Finally, Novak's theory of *human constructivism* (*24*) combines cognition (concepts and reasoning skills) with the affective domain (attitudes and motivation) and the psychomotor domain (dexterities and precision) (*25*).

Constructivist methods of teaching require that teachers, firstly, recognize their students' alternative ideas, and secondly take them into account in planning and performing their teaching, so that the aim of *conceptual change* is fulfilled. Conceptual change consists in the 'replacement' or 'substitution' of misconceptions with the corresponding scientific concepts. Research has shown that such a change is very hard to accomplish. Students, even if they come close to realizing the errors in their established thinking, revert very easily to their previous ideas, with which they are more comfortable (*13, 26*). It cannot be achieved by traditional didactic methodology, but through active, constructivist approaches. In particular, *concept addition* and *concept modification* toward the scientific option are the proper actions and better terms than 'concept replacement/substitution'. Strike and Posner (*27*) listed four conditions for bringing about conceptual change: (i) dissatisfaction (*dissonance*) with an existing conception; (ii) minimal understanding of the new conception (a person must realize how the new conception can restructure experience); (iii) the new conception must have the capacity to solve problems that the old one could not; (iv) the new conception must be fruitful, opening up new areas of thinking and learning.

Certainly most theories are neither complete nor perfect. They need constant revision and improvement. They may even need replacement. Even if various theories might appear as conflicting, there is benefit from an exposure to and use of the successes of each one of them (*28*). Adey (*29*) refused to view Piagetian theory and the alternative conceptions movement as irreconcilable rivals and thought it very likely that they will eventually be combined. In any case, it is widely accepted that constructivism "has done a service to science and mathematics education by re-emphasizing the importance of prior learning... by stressing the importance of understanding as a goal of science instruction, by fostering pupil engagement in lessons, and other such progressive matters (*19*)." There is a warning, however, against the idea that constructivism has all the answers. A major argument of critics of contructivism, as applied to science education, is that contructivists pay attention to how students learn (how they construct their concepts), but not to what knowledge (wrong or correct) they construct (*19*).

Methods of Educational Research

Educational research utilizes research methods from the social sciences. These methods are broadly distinguished into quantitative and qualitative ones.

Quantitative methodology uses large or relatively large samples of subjects (as a rule students) and tests or questionnaires to which the subjects answer. Results are treated by statistical analysis, by means of a variety of parametric methods (when we have continuous data at the *interval* or at the *ratio* scale) or nonparametric methods (when we have *categorical* data at the *nominal* or at the *ordinal* scale) (*30*). Data are usually treated by standard commercial statistical packages. Tests and questionnaires have to satisfy the criteria for *content* and *construct validity* (this is analogous to lack of systematic errors in measurement), and for *reliability* (this controls for random errors) (*31*).

Qualitative methodology uses small samples of subjects and the method of (usually) personal interviews, based on structured or more usually semi-structured questionnaires. The interviews are tape- or video-recorded and then transcribed and analyzed for patterns and categories in students' thinking. Data treatment can also be done by commercial packages for qualitative analysis. This methodology provides the possibility for in-depth monitoring and study of students' ideas and understandings about scientific concepts. One could think of written questionnaires as instant pictures, and of interviews as motion pictures.

There are a number of different approaches to qualitative methodology. Phenomenography, for instance, explores people's different ways of describing their common experiences or understandings about phenomena in the world (*32*). For the phenomenographer, the individual is not the target of the analysis. The transcribed material produces undivided data or a 'pool of meanings' to be analyzed. The analysis leads to a hierarchically ordered set of categories of description out of the logical relationships that are found between the different ways of understanding.

Because of small samples, results from qualitative work cannot easily be generalized. Very important is then the integration of quantitative and qualitative methodologies (*33-35*). According to Tobin (*34*), "qualitative and quantitative data could contribute in complimentary ways to the solutions of problems." Finally, because of various uncontrolled factors, results of educational research are difficult to duplicate and generalize. One method that produces more reliable results is the combination of a number of similar studies through the method of *meta-analysis*. The following are criteria that justify the use of meta-analysis (*36*): (i) the research question should be the same or at least similar; (ii) the effect of interest should be measured using the same scale in each of the studies to be combined; and (iii) all studies should be of high quality.

Physical Chemistry and Science Education

The study of physical chemistry is of utmost importance in the training of chemists. On the other hand, it is perceived as a difficult course. According to Moore and Schwenz (*37*), this course "sets the tone" of the chemistry major. As its name implies, physical chemistry is very closely related to physics, and this differentiates it from other mainstream chemistry and is the cause of its added difficulty for chemistry students.

Physical chemistry and physics may be different fields but they have some important features in common: they are abstract; they both use mathematics; they overlap in some content areas (such as thermodynamics and quantum mechanics). To a large extent, science and physics educators started research on basic physics concepts that also are used in physical chemistry. Consequently, physical chemistry education research owns much to the work that has been done in physics education and has much in common with it. For example, they share some of the research methodology and an interest in studying the relationship between the physical description of phenomena and its mathematics description in the learner's mind.

The Role of Meaningful Learning and Conceptual Understanding

Physical chemistry involves abstract and complex concepts and processes, so learning is difficult without a thorough understanding of the subject. Otherwise, students have to resort to rote learning of definitions, formulas, and processes. According to Ausubel (38), meaningful learning is "a process that is considered qualitatively different from rote learning in terms of non-arbitrary and non-verbatim reproduction of the content that is to be learnt to existing ideas in cognitive structure. Both rote/meaningful and reception/discovery dimensions of learning exist on a continuum rather than being dichotomous in nature." Novak (39) makes explicit that "Meaningful learning at one edge of the continuum requires well organized relevant knowledge structure and high commitment to seek relationships between new and existing knowledge. Rote learning at the other edge results from little relevant knowledge poorly organized and little or no commitment to integrate new with existing relevant knowledge." Evidence of meaningful learning occurs when tests of comprehension are presented in a somewhat different context than those originally encountered. Novak shows how meaningful learning and the transfer of knowledge relate (39).

The extent and complexity of meanings we hold in any domain are dependent on the quality and quantity of meaningful learning we have pursued in that knowledge domain. In turn, the quantity and quality of the knowledge structures we build will determine our ability to transfer this knowledge for use in new contexts.

The Role of Models

According to Justi and Gilbert (40), "a model: (i) is a non-unique partial representation of an object, an event, a process or an idea; (ii) can be changed;

(iii) is used for enhancing visualization, as a way of both supporting creativity and favoring understanding, in making predictions about behavior or properties; and (iv) is accredited by adequate groups in society." Grosslight et al. (*41*) explored the variance in conceptualization of the notion of model, and identified three levels of understanding: level 1, models as replicas of reality; level 2, they serve a specific purpose, with the emphasis still on reality; level 3, models thought of not as replicas of reality but as constructions that serve specific purposes, one of which is to test ideas.

Models are broadly distinguished on the one hand into material or physical or concrete, and on the other hand into abstract or conceptual or symbolic. Symbolic models include mathematical formulas and equations. Gilbert et al. (*42*) classified models according to their ontological status as follows: mental, expressed, consensus, scientific, historical, and hybrid. Consensus, scientific, and historical models specifically developed for education are distinguished into teaching, curricular, and models of pedagogy.

Models play an important role in physical chemistry education. Research has shown that students at all levels prefer concrete or simple abstract models, for example: space-filling models of atoms and molecules (*43*); the Bohr model of the atom (*44-46*); or the octet rule (*47*). These preliminary models are very stable. Although students at a higher educational level may have been exposed to abstract models of higher explanatory power, such as the quantum mechanical model of the atom or the theory of molecular orbitals, they find it difficult to replace the earlier models. What is frequently observed is that learners accommodate new knowledge into preexisting knowledge, constructing personal meanings and alternative mental models that contain elements from all earlier models (hybrid models) (*48*).

Factors that Are Involved in Learning Physical Chemistry

Nicoll and Francisco (*49*) carried out a correlational study on the effect of a variety of factors that may influence success in physical chemistry. The results from the students' ($n = 77$) and teachers' ($n = 47$) surveys indicated a clear disparity. The interpretation of the findings was that students and professors had different perceptions about what factors influence students' performance in the course. The authors reported significant relationships between students' grades and their logical thinking skills, but not with their information processing ability (see below), their previous chemistry knowledge, the extent of their exposure to mathematics, or their attitudes toward the course. No single predictor of performance (in mid-term and final semester exams) was found. While basic mathematical ability was an important factor, mathematical skills alone were not the best predictor of performance. Moreover, the numbers of mathematics courses the students took before physical chemistry had no correlation with actual performance in physical chemistry.

Derrick and Derrick (*50*) also examined factors that are considered as predictors of success in physical chemistry. Chemistry, physics, and mathematics grades and the number of times these courses were repeated were collected for 60 physical chemistry students over a long period (1976-1999). Contrary to the findings of Nicoll and Francisco, they reported that success in prior mathematics courses affected positively physical chemistry grades. Success in calculus (but not in college algebra) was found to have a strongly significant effect. Success in previous chemistry and physics courses correlated also positively to success in physical chemistry.

Hahn and Polik (*51*) investigated further various factors that affect success in physical chemistry, with the aim to corroborate and expand upon the factors in the Nicoll and Francisco study. They studied two physical chemistry courses using a larger student data set ($n = 279$). Motivation and study skills were two added factors in the study. To improve reliability of data, they also used student transcripts. Measures of success in physical chemistry were free-response exams written by the course instructor given during the semester, standardized final exam from the ACS DivCHED Examinations Institute, and the final course grade. As in the case of Derrick and Derrick, the findings disagreed with those of Nicoll and Francisco regarding mathematical ability, reporting that mathematical skills play an important role. They attributed this to mathematical techniques being a central part of solving problems in physical chemistry. The results of the correlational study indicated that mathematics performance, homework scores, and number of mathematics courses taken were all important factors. They also related to the four factors perceived by professors as most important to success in physical chemistry in the Nicoll and Franciso study (basic mathematical skills, logical thinking skills, motivation, and study skills).

Finally, Sözbilir (*52*) studied Turkish undergraduate students' and lecturers' perceptions of students' learning difficulties in physical chemistry. Data were collected from the chemistry departments in two universities ($n = 47$ and 44 respectively). Students worked in pairs or in groups of three, and answered a free-response survey consisting of two questions, at the end of the sixth semester after having been taught physical chemistry. In addition, the two instructors of the course in the two universities were interviewed. On a number of issues there was agreement between students and lecturers: the abstract nature of the concepts of physical chemistry; the overloaded course content; insufficient resources; teacher-centered, expository teaching; lack of student motivation. On the other hand, there were complaints by the students about having to memorize definitions and facts, the lack of relation among lectures, labs, and exams, and the high-level of mathematics used while less attention is paid to the concepts. The lack of links between the concepts of physical chemistry and their applications to industry and everyday life, as well as with their students' future careers, were also judged by the students as contributing to lack of motivation. Finally, lecturers were critical of the overcrowded classes and the lack of staff.

Turning to the implications from the above studies, Nicoll and Francisco (*49*) advised instructors to weigh the various factors that may influence success in physical chemistry, and to be careful in incorporating new knowledge into the physical chemistry curriculum. Derrick and Derrick (*50*) considered that small classes are very effective, making considerable attention possible for students and the opportunity for early intervention when the students have difficulty with the material. Hahn and Polik (*51*) recommended regular assignment of homework, communication to the students what they should have done so that to encourage good study skills, and assignment of homework problems and questions that are similar to those appearing in the course examinations. Sözbilir (*52*) urged that constructivist methods of teaching, that play more attention to quality than quantity of content, should be employed; in addition, there is a need for a flexible curriculum that will take into account the rate of students' learning and needs. Finally, it is the instructor's role to coordinate the four variables that influence learning, namely, characteristics of the learner, nature of learning activities, nature of assessment, and characteristics of materials (*52, 53*).

Mathematics and Physical Chemistry

Mathematical operations occur regularly both in the theory and problems of physical chemistry, contributing greatly to the complexity of the subject. Mathematics is essential for the meaningful learning of physical chemistry, but for this to happen it must be coupled with understanding of the underlying physical concepts.

In the Nicoll and Francisco study (*49*), a math diagnostic was used that had been designed by the professor of the course to assess the math skills most commonly encountered in physical chemistry. It consisted of ten questions: 6 on calculus, 2 on algebra, and 2 on word problems. Word problems (for instance a simple problem in solution chemistry) required no knowledge of either chemistry or higher mathematics in order to solve them. It was found, that the ability to solve such word problems correlated most significantly with course grade. On the other hand, the ability to perform calculus-based problems correlated statistically significant but weaker. Also, as stated above, the number of mathematical courses taken had no bearing on their performance in physical chemistry.

Derrick and Derrik's students (*50*) who had earned A and B in physical chemistry had been highly motivated and possessed both strong mathematics and strong cognitive skills. Students who had earned C were either less motivated or weak in mathematics or cognitive skills. In the Hahn and Polick study the average grade in the mathematics courses received the highest correlation coefficients in all three measures of success in physical chemistry (Pearson's r = 0.72, 0.50, and 0.72 respectively). The number of mathematics courses taken did not correlate as highly as the average mathematics grade, suggesting that mathematical ability is more important than exposure. Note that the authors attributed the discrepancy between their finding and that of Nicoll and Francisco to the different methodology used in the student reporting of the number of mathematics courses taken.

Several older studies (54-57) have attempted to correlate mathematical skill and student reasoning ability with success in physics, which is a prerequisite for, and is of a similar nature with physical chemistry. Mathematical skills seem to be necessary but not sufficient for success in physics. There are students with marginal mathematical skills, but with well-developed logical and conceptual skills who can be successful in physics.

Apparently, different instructors may place different demands on the students with regard to mathematical ability. Some may be content with the capacity of students to connect physical chemistry with mathematics, while others may pay more attention to mathematical operations and calculations. These two distinct approaches may lead to different outputs and attitudes.

Educational Research in
Various Subfields of Physical Chemistry

Next, we review findings of educational research about the main areas of physical chemistry. Most of the work done was in the areas of basic thermodynamics and electrochemistry, and some work on quantum chemistry. Other areas, such as chemical kinetics, statistical thermodynamics, and spectroscopy, have not so far received attention (although the statistical interpretation of entropy is treated in studies on the concepts of thermodynamics). Because many of the basics of physical chemistry are included in first-year general and inorganic courses (and some even in senior high school), many of the investigations have been carried out at these levels.

Thermodynamics

Much of the research that has been done to date on learning in the domain of thermodynamics has involved the study of populations other than college age physical chemistry students. Thermodynamics is usually first introduced in elementary and introductory physics courses. For this reason, many studies have been conducted from the physics education perspective (56-60). These studies provide insights into the learning of important basic concepts, uncovering misconceptions that students bring into physical chemistry courses.

Students have severe difficulties in differentiating between heat and temperature. The distinction between extensive and intensive properties is crucial for making the differentiation of heat from temperature (60). In addition, students find it difficult to apply the particle model to explain thermal processes. The meaning of the term 'energy' is another cause of considerable confusion. Heat is closely related to energy, with heat energy often considered only as a cause of temperature changes, and motion only as an effect of kinetic energy. Students fail to see temperature-equalization processes as caused by the interaction between the hot and the cold body. In addition, many students do not think that the temperature-equalization process may be spontaneously reversed (that is, temperature differences may occur by themselves). Kesidou & Duit (60) concluded that a totally new teaching approach to heat, temperature, and energy is necessary for a proper understanding of the second law of thermodynamics. In this approach, basic qualitative ideas of the second law should be a central and integral part from the beginning of the instruction.

Looking at thermodynamics from the physical chemistry perspective, students at the high school and college levels experience difficulties with fundamental concepts in chemical equilibrium and thermodynamics (61). Thomas and Schwenz (62) found that physical chemistry students still have difficulties with the above concepts, which may continue through their professional careers. Both students and lecturers in the Sözbilir study (52) (see above) assumed the abstract nature of thermodynamics concepts as a cause of learning difficulties.

Carson and Watson (63) have studied UK first year university students' understanding of enthalpy changes. Ten statements covering the knowledge and understanding about enthalpy changes were drawn up, and students' answers were compared to these. Most students held the same conception about enthalpy as pre-university students studying chemistry at advanced (A-) level. They did not see the meaning of enthalpy as problematic. Problems were found with the notion of pV work. Also, they did not understand the meaning of expressions like $\Delta H = \Delta U + p\,\Delta V$. The use of terms for heat and work that have different scientific meaning from that in everyday usage was assumed to cause many of the problems.

Thomas and Schwenz (62) investigated concepts of equilibrium and fundamental thermodynamics at the physical chemistry level: the first law of thermodynamics and related topics; the second law and entropy, spontaneous change and Gibbs energy; equilibrium, factors affecting equilibrium composition, and factors determining equilibrium constants. The study identified, classified, and characterized students' conceptions in comparison with those of experts, as expressed in textbooks. The results made it clear that students still had difficulties with chemical equilibrium, and pointed at the difficulty lecturers experience with attempting to modify students' conceptions through standard instruction. Thirty alternative conceptions and non-conceptions were identified, occurring in ranges from 25 to 100% of students (average: 52%).

In another study, Carson and Watson (*64*) aimed at exploring further the qualitative understanding of students. A set of nine scientifically accepted statements about the concepts of entropy and free energy were developed, to which students' answers were compared. It was confirmed that chemistry students had formed some misconceptions related to thermodynamics before they began their university studies, and these had a significant impact on their understanding. Confusion in the use of thermodynamics terms like enthalpy, energy, entropy, and kinetic energy (also reported in previous studies) was found. Entropy was described in vague terms such as chaos or randomness. Energy changes were related to changes of state rather than to distribution of energy in microstates. 'Forms of energy' was added as a persistent alternative framework that interferes with new concepts in thermodynamics.

Gibbs free energy was the subject of another study by Sözbilir (*65*) with Turkish chemistry undergraduate students. Open-ended diagnostic questions and interviews were used before and after the topic was taught. The study confirmed previous research findings about misconceptions and confusion of energy, enthalpy and entropy. In addition, it identified further misunderstandings. A lack of understanding of related concepts, such as equilibrium and reaction dynamics, energy, energy transformations, and the energy change involved in chemical reactions, was assumed as the origin of many misunderstandings.

Turning to recommendations, Carson and Watson (*64*) were critical of the emphasis given during lecture courses, example classes, and examinations to numerical calculations using thermodynamic equations, while no effort is made to elicit and promote students' qualitative conceptual understanding. They recommended that thermodynamic entities should be defined qualitatively and their effect discussed before they are defined and treated quantitatively. Furthermore, the essence of thermodynamics is a study of interactions. Entropy and free energy cannot be understood as isolated entities that can be transformed into one another. Students need to understand them in the context of chemical 'processes'/reactions. Sözbilir (*65*) argued that care should be taken to establish a secure knowledge of fundamental chemical concepts before teaching advanced ideas, for instance the difference between H and ΔH or G and ΔG. At the college general chemistry level, Teichert and Stacy (*66*) explored the effectiveness of intervention discussion sessions, in which the implications of research on student preconceptions, knowledge integration, and student explanation were taken into account. Two interventions, one on bonding and one on spontaneity and free energy change, were tested. It was found that the intervention group outperformed a control group.

The various "canonical" representations of thermodynamics are an issue of concern. Tarsitant and Vicentini (*67*) analyzed a number of textbooks on thermodynamics, and pointed out divergent attitudes not only towards the definition of fundamental thermodynamic concepts, but also towards the epistemological status of the subject. According to these authors, these attitudes underlie the relationship between the macroscopic/phenomenological and the submicroscopic/statistical approach on one side and the 'state' or 'process' approach on the other. Important for a deeper understanding of these mental representations is their historical reconstruction and epistemological analysis. Finally, De Berg (68) studied the mathematical analogy between van't Hoff's law for osmotic pressure and the ideal gas law from two points of view: as an artifact of the mathematical thermodynamic treatment of the phenomenon of osmosis, and as a result of a controversial kinetic molecular model that is assumed to have more educational value than the thermodynamic model.

Electrochemistry

Electrochemistry is a topic introduced in high school and first year general and analytical chemistry courses. Research has shown that many of the problems associated with electrochemistry are caused at earlier stages of education. A good knowledge of redox reactions is a prerequisite for the study of electrochemistry, but this topic presents its own difficulties (69). Garnett and Treagust (70, 71) studied conceptual difficulties by senior high school students in two areas: (i) electric circuits and oxidation-reduction equations; (ii) galvanic and electrolytic cells. They examined several senior chemistry textbooks for relevant conceptual and prepositional knowledge, and used this information to formulate knowledge statements that were included in the interview protocol. Several students were confused about the nature of electric current both in metallic conductors and in electrolytes. They also experienced problems in identifying oxidation-reduction equations. Misconceptions were also revealed in relation to the sign of the anode and cathode. Students who thought the anode was negatively charged believed cations would move toward it, while those who thought it was positively charged could not explain why electrons move away from it. In the case of electrolytic cells, many students did not associate position of the anode and cathode with the polarity of the applied electromotive force (emf). Confusion between galvanic and electrolytic cells was also detected. Finally, the authors incorporated eight misconceptions identified into an alternative framework about electric current, grounded on the notion that a current always involves drifting electrons, even in solution.

According to Birss and Truax (72), students are likely to experience confusion and difficulty with more advanced treatments of the subject. With regard to conceptual difficulties, the authors looked at the equilibrium potential, the reversal of sign of electrode reactions that are written as oxidations, and the differences between galvanic (electrochemical) and electrolytic cells. An approach for teaching these topics at the freshman level was then proposed. In this approach, concepts from thermodynamics and chemical kinetics are interwoven with those of electrochemical measurements. Very useful are schematic plots of current versus potential in a galvanic cell without or with current flow.

Ogude and Bradley (73) investigated pre-college and college students' difficulties regarding the qualitative interpretation of the submicroscopic processes that take place in operating galvanic cells. Four areas were identified: conduction in the electrolyte; electrical neutrality; electrode process and terminology; and aspects relating to cell components, current, and cell emf. A questionnaire was used to determine how widespread the misunderstandings identified in the four areas above were at different educational levels and to determine their possible causes. According to the findings, the relevant misconceptions were present among students both in high school and at the tertiary level. Traditional methods of teaching and some textbooks were held responsible for causing or fostering some of the misconceptions by paying attention to quantitative/manipulative aspects of electrochemistry. Very little concern is paid with regard to qualitative understanding of the concepts and the processes involved.

Sanger and Greenbowe (74) replicated the research done by Garnet and Treagust with modifications and expanded it to include concentration cells. Most commonly encountered misconceptions included: notions that electrons flow through the salt bridge and electrolyte solutions; that the plus and minus signs assigned to the electrodes represent net electronic charges; that water is unreactive in the electrolysis of aqueous solutions; and that half-cell potentials can be used to predict the spontaneity of individual half-cells, while galvanic cell potentials are independent of ion concentrations. The study added evidence showing that most students demonstrating misconceptions were still able to deal successfully with algorithmic quantitative problems. The authors considered unawareness of the relative nature of electrochemical potentials, and misleading and incorrect statements of textbooks as probable causes of the misconceptions.

Sanger and Grenbowe (75) combined the data from their earlier study with those of Garnett and Treagust, and compiled a list of students' common misconceptions in electrochemistry. The list covered galvanic cells, electrolytic cells, and concentration cells. The role attributed to misleading or erroneous statements in textbooks as sources of misconceptions led these authors to analyze a number of general chemistry textbooks (75).

Özkaya (76) studied conceptual difficulties experienced by prospective teachers in a number of electrochemical concepts, namely half-cell potential, cell potential, and chemical and electrochemical equilibrium in galvanic cells. The study identified common misconceptions among student teachers from different countries and different levels of electrochemistry. Misconceptions were also identified in relation to chemical equilibrium, electrochemical equilibrium, and the instrumental requirements for the measurement of cell potentials. Learning difficulties were attributed mainly to failure of students to acquire adequate conceptual understanding, and the insufficient explanation of the relevant concepts in textbooks. Özkaya et al. (*77*) worked further with prospective teachers and studied their understanding of galvanic and electrolytic cells. The study identified new misconceptions in addition to known ones.

Finally, Niaz (*78*) designed a teaching strategy that can contribute to improvement in performance in basic electrochemical calculations (balancing redox equations, using standard electrode potential values for predicting the direction of a cell reaction, and applying the laws of electrolysis). The strategy used 'teaching experiments' that provided students with both the correct and alternative responses to problems. In this way, a conflict situation was created, leading to conceptual change. The results of this study with freshman students at a major university in Latin America showed statistically significant difference in favor of the experimental group.

The numerous concepts and aspects involved in electrochemistry (conduction in the electrolyte, electrical neutrality, electrode processes, polarity of electrodes, cell potentials, electrochemical equilibriums, the role of salt bridges, etc.) are complex and interwoven, and cannot be understood in isolation from each other; instead, the electrochemical cell should be understood in its entirety (*73*). Garnet and Treagust (*71*) recommended that teachers and textbook writers need to be cognizant of the relationship between physics and chemistry teaching, of the need to test for erroneous preconceptions about electric current before teaching galvanic and electrolytic cells, and of the difficulties experienced by the students when using more than one model to explain scientific phenomena.

Sanger and Grenbowe (*75*) used their textbook analysis for drawing a list of five general suggestions for instructors and textbook authors: (i) avoid the use of simplifications (such as always drawing the anode as the left-hand half-cell); (ii) avoid the use of vague or misleading statements and terminology (such as "ionic charge carriers"); (iii) calculate cell potentials using the difference method ($E^\circ_{cell} = E^\circ_{cathode} - E^\circ_{anode}$) instead of the additive method ($E^\circ_{cell} = E^\circ_{ox} + E^\circ_{red}$); (iv) avoid using simple electrostatic arguments to predict ion and electron flow in electrochemical cells; (v) always consider all possible oxidation and reduction half reactions, including the reactions of the electrodes, water and aqueous species.

Finally, promising for the overcoming of relevant misconceptions is the coupling of teaching with computers (*79, 80*). In particular, animations appear to be helpful in visualizing chemical processes on the molecular level. Computer animations and simulations are most effective when coupled with actual demonstrations or working in the laboratory with electrochemical cells (*80*).

Quantum Chemistry

Elementary quantum-chemical theories of atomic and molecular structure [including atomic orbitals (AO) and molecular orbitals (MO)] are part of the upper secondary curriculum in many countries. They are also taught in general chemistry or introductory inorganic chemistry courses in chemistry and other science departments.

Numerous misconceptions occurring with students at the high school or the pre-physical chemistry tertiary level have been identified. Harrison and Treagust (*81*) reported that senior high school students confused electron shells and electron clouds. Taber (*82, 83*) found that British advanced high school (A-level) students had real difficulties making sense of orbital ideas, treating the terms orbitals, shells, and orbits, interchangeably. They also became confused between the mathematical modelling (LCAO) of MO formation, and the orbitals themselves, referring to 'linear orbitals'. Given that, it is not surprising that they did not readily develop the concepts of MOs: "as an appreciation of MOs is built upon an understanding of the simple atomic case, it is to be expected that attempting to teach the more complex examples whilst students have limited conceptualizations of the simpler case may only compound their difficulties." Taber (*84*) extended his previous studies by drawing a 'typology of learning impediments.' The typology can be used by the instructors for diagnosing the origins of students' difficulties in learning orbital ideas.

Tsaparlis and Papaphotis (*85*) reported that twelfth-grade students did not have a clear understanding of orbitals, among other misconceptions; for many, the orbitals represented a definite, well-bounded space; also, they did not realize the approximate nature of AOs for many-electron atoms. Zoller (*86*) argued that misconceptions and misunderstandings can develop among students in learning hybridization because of the problems related to understanding the meaning of fundamental concepts involved, such as the concept of atomic orbital and the real meaning of the s, p, d, and f orbitals. Nakiboglu (*87*) also found in her students serious misconceptions about hybridization, arising from problems with quite important pre-requisite knowledge, especially with the concept of atomic orbital. Coll and Treagust (*88, 89*) examined the advanced (upper secondary, undergraduate, and graduate) students' mental models of chemical bonding, and found that despite the participants' competence in the description and use of more abstract models (especially when simple models had inadequate explanatory power), all these learners (including MS and PhD students) preferred simple, realistic models, and related to more abstract models only in the context of tests or examinations. Furthermore, the students struggled to use their mental models to explain the physical properties of covalently bonded substances.

On a more physics-oriented approach to quantum mechanics, Kalkanis et al. (*90*) accepted that the main misconceptions are caused by the overlapping/mix-up of the conceptual frameworks of classical and quantum physics, and from epistemological obstacles to the acquisition of the proper knowledge. Furthermore, they proposed an educational strategy for a simple, qualitative and sufficient approach to quantum mechanics by prospective teachers. The strategy aims at a conceptual structure that includes classical and quantum physics as two totally independent systems. The complete distinction of the two systems demands a radical reconstruction of students' initial knowledge that is based on the juxtaposition of the two models. Greca and Freire (*91*) have chosen a didactic strategy that puts the emphasis on the quantum features of the systems, instead of searching for classical analogies. In particular, the method considers the concept of quantum state as the key concept of quantum theory, representing the physical reality of the system, independent of measurement processes. More than half of the students involved in the implementation of the strategy attained a reasonable understanding of the basics of quantum mechanics.

It follows from the above that students arrive at the quantum chemistry course carrying with them from previous instruction a number of misconceptions and incomplete knowledge about quantum-chemical concepts. This is attributed mainly to the elementary, imprecise and mostly pictorial coverage of the relevant concepts. Shiland (*92*) analyzed a number of secondary chemistry texts and found that they were not satisfactory with respect to the presence of the following four elements associated with a conceptual change model: dissatisfaction, intelligibility, plausibility, and fruitfulness. The conceptual change should aim at the students accepting the better rationale of quantum mechanics over simpler atomic models such as the Bohr theory (a very difficult task).

A basic question is whether chemistry graduates have a deep and precise understanding of modern concepts of AOs, MOs and related concepts. An analysis of examination data of students who had passed the compulsory quantum-chemistry course was carried out by Tsaparlis (*93*). Most students failed to provide an exact definition for an AO, such as "a one-electron, well-behaved function that can describe - more or less successfully - the behavior of an electron in an atom" (and similarly for an MO). For some students, an AO was understood as or connected with "a region in space inside which there exists a given probability, for example 90%, for an electron to be encountered." Also, a significant proportion of the students identified an MO only with a linear combination of AOs. Another source of confusion was the fact that the actual solutions of the Schrödinger equation are complex functions, except for the *s*-type orbitals. Impressive was the misinterpretation of the figure eight '*p*-type atomic orbital,' familiar from previous instruction - this is a cross-section of the graph of the squared spherical harmonic, $Y^2(\theta, \varphi)$, for the p_z AO; it *does not* give the shape of a p_z orbital. Very few students recognised the *equal-probability contour* (or *boundary surface*) for a p_y orbital.

In many-electron atoms, the Schrödinger equation cannot be solved exactly, so approximations must be made. The simplest and crudest approximation is to neglect entirely electron-electron interactions (repulsions) and electron spin. In this way, hydrogenic orbitals are found as solutions. Into these orbitals we then place the electrons, according to the *aufbau principle*, and thus derive electron configurations. More sophisticated methods are available that take into account, in an approximate fashion, the electron-electron interactions. All these involved 'details' that do not become knowledge for many students. The concept of Slater determinants, their definition, and the approximations involved proved difficult. On the other hand, writing all Slater determinants arising from a given electron configuration (an algorithmic process) was not difficult. Similar difficulty appeared for the concepts of spectroscopic terms, while finding the term symbols for a given configuration proved again an easy task.

Pauling and Wilson (94) stated over seventy years ago that

"Quantum mechanics is essentially mathematical in character, and an understanding of the subject without a thorough knowledge of the mathematical methods involved and the results of their application cannot be obtained."

And as Coulson (95) has put it:

"Mathematics is now so central, so much 'inside', that without it we cannot hope to understand our chemistry ... These (quantum-chemical) concepts have their origin in the bringing together of mathematics and chemistry."

It is then quite understandable why, without the necessary mathematical machinery, the relevant concepts cannot be properly grasped. On the other hand, the mathematical disguise that is characteristic of quantum-chemistry courses makes both teachers and students pay more attention to the complexities of the mathematics (the tools, the 'trees') and lose the physics (the actual world, the 'forest'). Although mathematics is essential for a deep understanding of quantum chemistry, the underlying physical picture and its connection with mathematics are equally important. AOs, MOs and related concepts derive from Schrödinger's wave mechanics, which is an approximation to nature. According to Simons (96), "orbital concepts are merely aspects of the best presently available model; they are not 'real' in the same sense that experimental observations are."

The physics of quantum chemistry is complicated and different from classical physics. It has been argued that thinking abilities beyond Piagetian formal operations may be of major importance for an adequate understanding of quantum-mechanical (and relativistic) issues (97). These *post-formal operations* include what has been termed as *quantum logic* (98). Although one can derive the Schrödinger equation with entirely classical arguments (99) (with Planck's constant h serving as the bridge between classical and quantum mechanics), one has to admit that quantum mechanics has brought a new way of thinking about the physical world at the subatomic level. Karakostas and Hadzidaki (100) argued that a *realist* (as opposed to an *empiricist/constructivist*) interpretation of quantum theory is necessary for revealing the inner meaning of the theory's scientific content; according to these authors, this requires the abandonment or radical revision of the classical conception of physical reality.

A feature of most introductions to quantum chemistry is their postulative approach. Although the Schrödinger equation cannot be proved or derived strictly, there are many ways to introduce this equation that provide insights into the meaning of quantum mechanics. Tsaparlis (*101*) has suggested an approach from the historical perspective, studying the methods of the pioneers Schrödinger, Heisenberg, and Dirac; in addition, he made a synthesis of various modern heuristic treatments into a coherent and meaningful whole.

Quantum theory suggests that, strictly speaking, "atomic orbitals can no longer be said physically to 'exist' in anything except one-electron systems; many-electron orbitals are ontologically redundant" (*102*). And yet, we know that chemists are very comfortable in using orbitals everywhere in a quasi-classical manner that is judged to be in conflict with the essence of quantum mechanics (*103*). The extensive use of electronic configurations of atoms in chemistry textbooks (with a lot of relevant practice questions), reinforces further the impression about the fundamental nature of orbitals and configurations (*104*). This misconception is also extended by the modern visualisation of orbitals by means of computers (*105, 106*). We must be aware that the way chemists see and use quantum mechanics is essentially different from that of physicists (*103, 104*), with the result that the need for a 'philosophy of chemistry' has arisen (*104*). Accordingly, the view that chemistry has been reduced to physics, or more specifically quantum mechanics, is mistaken, according to Scerri (*102*).

Finally, we must recognize that the complexity of quantum chemistry concepts, theories, and approximations is such that it will be difficult to expect that instruction could leave no lingering misconceptions (zero error tolerance). However, instructors should be aware of the problems, and have themselves a deep knowledge and understanding of all 'variables', methods, and issues.

Problem Solving and Laboratory Work

Problem Solving in Physical Chemistry

Problem solving is an important and integral part of physical chemistry in addition to the concepts, principles and methods. There is a vast range of problems: closed problems, with one answer; open problems, which can have more than one answer and for which data may not be supplied; problems that can be solved by pencil-and-paper or by the computer; problems that need experiment in order to be solved; and real-life problems versus scientific problems or even thought problems. A thorough classification of problem types has been made by Johnstone (*107*).

It is accepted that physical chemistry instructors must give their students ample practice on straightforward problems (that is, exercises) that would give them familiarity with the application of the equations of physical chemistry. On the other hand, according to Ritchie et al. (*108*), it is very important to provide problems with which "students are asked to apply chemical reasoning to something approaching a real-life situation;" further, "it is vitally important for students to get into the habit of looking up facts and figures for themselves." For this reason, not all the required data are written into such problems.

It follows from the above that a distinction must be made between problems and exercises, with the latter requiring for their solution only the application of well-known and practiced procedures (*algorithms*). The skills that are necessary for the solution of exercises are, as a rule, *lower-order cognitive skills* (LOCS). On the other hand, a real/novel problem requires that the solver must be able to use what have been termed as higher-order cognitive skills (HOCS) (*109, 110*).

In the earlier studies of problem solving in the physical sciences, attention was centered on the differences in the methods and procedures used by students and by experts when solving problems (*111-113*). The basic differences were: (a) the comprehensive and complete scheme of the experts in contrast to the sketchy one of the novices; and (b) the extra step of the qualitative analysis taken by the experts, before they move into detailed and quantitative means of solution.

The early stages in problem solving involve understanding the problem, that is building an *internal* (or *mental*) *representation* of it. In Larkin's words (*114*), "To work on the problem, the solver must convert the string of words with which he is presented into some internal mental representation that can be manipulated in efforts to solve the problem. Understanding the problem then means constructing for it one of these internal representations." Bodner and coworkers (*115, 116*) studied the correlation of spatial ability with performance in general chemistry and organic chemistry college courses. The tests of spatial ability used were tests of disembedding (see below) and cognitive restructuring in the spatial domain. These tests correlated best with the students' performance on novel problems, rather than algorithmic exercises. The conclusion was that very important factors for success in problem solving are the early preliminary stages in problem solving that involve disembedding the relevant information from the statement of the problem and restructuring or transforming the problem into one the individual understands.

Tsaparlis (*117*) carried out a correlation study of the role of various cognitive factors (variables) in the solution of non-algorithmic quantitative problems in elementary physical chemistry. The cognitive variables were: scientific reasoning (developmental level), working-memory capacity; functional mental capacity (M-capacity); and disembedding ability.

Scientific reasoning is a measure of a student's level of intellectual development (*118*). [In previous work of various authors, the term 'developmental level' (which is associated with Piaget's developmental theory) has been used instead.]

Working memory refers to the human limited capacity system, which provides both information storage and processing functions (*119-121*), and is necessary for complex cognitive tasks, such as learning, reasoning, language comprehension, and problem solving. Related to the working-memory capacity is the *mental capacity* or *M*-capacity, which derives from Pascual-Leone's theory of Constructive Operators (*122-124*). *M*-capacity or *M*-space represents "a reserve of mental energy that can be allocated to raise the activation weight of task-relevant schemes." *M*-capacity, is further distinguished into the maximum available *M*-capacity or structural *M*-capacity (M_s), and the actually mobilized *M*-capacity or functional *M*-capacity (M_f). Both working-memory capacity and *M*-capacity refer to information holding and processing capacity.

Disembedding ability refers to the degree of field dependence/field independence, and represents the ability of a subject to disembed information in a variety of complex and potentially misleading instructional contexts (*125, 126*). This ability is also connected with the ability of the subject to separate *signal* from *noise*; thus, learners who use some of their memory capacity to process irrelevant data ('noise') appear to posses lower working-memory capacity, and are categorized as field dependent.

Seven separate studies were carried out with chemistry students taking basic physical chemistry courses (*117*). Students had to solve a novel problem in an open-book, end-of-semester examination. Seven problems were used, mostly taken from the book by Ritchie et al. (*108*). One of the problems (adapted from Ritchie et al.) is reproduced below:

A chemist in Mexico city, where the atmospheric pressure is 7.81×10^4 Pa (that is $p/p^\circ = 7.81 \times 10^4/10^5 = 0.781$ bar), determined the *pH* of a solution at 25.0°C by making it the solution in a hydrogen electrode and measuring the potential of the electrode relative to some reference electrode (a calomel electrode). In calculating the *pH* of the solution from the potential measurement, he assumed, erroneously, that the (hydrogen) gas bubbling out of the hydrogen electrode was at a pressure of 1 bar. If he found the *pH* to be 5.00, what was the correct value?

As a rule, students were not supplied with necessary data (facts, figures, values of constants, etc.) for solving the problems, but they had to search for them in their textbooks. In addition, the students were tested in a follow-up, closed book examination whether they had acquired the necessary knowledge (the partial steps) for solving the corresponding open-book problem.

The problems used proved very difficult for the students. They are clearly not straightforward algorithmic problems and have a number of features of true/realistic problems. One could even think of extremely difficult problems combining conceptual understanding and physicochemical calculations (*127*).

The results of the seven studies were analyzed using the statistical method of *meta-analysis*. Homogeneity of the studies was checked with an appropriate statistical test, while combined Pearson correlation coefficients were estimated with four different methods. The main findings were:

- scientific reasoning (developmental level) showed lack of correlation;
- working memory showed weak correlations, but stronger than scientific reasoning;
- both functional M-capacity and disembedding ability played a very important role (in terms of correlation) in the solution of the problems.

Turning to the implications of the study, we first must take into account that, as far is known, working memory is genetically fixed. What must be increased is the individual student's efficiency in using whatever space he/she has. Part of this is disembedding ability; also 'chunking' ability. To our knowledge there are no established general methods or intervention programs that can be effective in increasing working-memory capacity or disembedding ability. However, practice in dealing with complex field situations and in using 'chunking devices' could increase efficiency.

Field dependence-independence involves not just perceptual field, but also a degree of information processing (*128*). It has been reported that a field independent but low information-processing capacity student should demonstrate in problem solving similar cognitive behavior with a field dependent but high information-processing capacity student (*128-130*). Field-dependent students have specific requirements as regards teaching methods, so teaching materials require careful attention (*131*). Note that the distribution of students into the three levels of disembedding ability usually follows (approximately) the normal distribution.

In conclusion, problem solving refers to students' ability to use their general background knowledge and apply it to novel situations. Information processing ability and disembedding ability are crucial in problem solving. Similarly, scientific reasoning and logical thinking skills (students' ability to apply logic and reason their way through the problem situation) certainly contribute. On the other hand, proficiency in mathematics plays often a vital role in physical chemistry problem solving. This proficiency requires specific content knowledge of how to do specific mathematical operations, such as integration, partial differentiation, solution of differential equations, or matrix algebra. According to Nicoll and Francisco (*49*): "Theoretically students should use all these skills while attempting to solve a physical chemistry problem. However, they are distinct skills that students acquire throughout their educational careers and use to varying degrees depending on the setting" (and their experience).

Taking into account the importance of the preliminary stages in problem solving (understanding the problem), Bodner and Domin (*132*) recommended encouraging students to use multiple representations when solving a problem. Students may "become more successful if we can convince them of the limitations of being trapped in a verbal/linguistic representation system." Symbolic/pictorial representations can provide alternative useful insights. Zoller and Tsaparlis (*110, 133*) made suggestions that facilitate the transition from lower-order cognitive skills (LOCS) to the desired higher-order cognitive skills (HOCS). Combined HOCS and LOCS-type, formal and informal, examinations and tests are needed for challenging and fostering students to develop their HOCS capacity. Finally, according to Frazer (*134*), problems in university courses must be challenging and real problems, selected from the chemical literature, with solutions that will not be obvious to the students, even though the necessary information and reasoning is likely to be within their grasp.

Laboratory Work

An integral part of a student's education in physical chemistry is laboratory/practical work. While it is generally accepted that the main purposes of laboratory work are to teach hand skills and to illustrate theory, significant problems have been identified in the science education literature about the laboratory courses, and in particular about the ineffectiveness of laboratory instruction in enhancing conceptual understanding (*135, 136*), and unrealistic in its portrayal of scientific experimentation (*137*).

Domin (*138*) distinguished four types of laboratory instruction: *expository, inquiry, discovery,* and *problem-based.* These styles can be differentiated by their outcome, their approach, and their procedure. Expository and problem-based activities typically follow a deductive approach, while inquiry and discovery activities are inductive.

The most commonly applied style of laboratory instruction in chemistry, including physical chemistry, is the expository one, which is instructor-centered. The learner has only to follow the teacher's instructions or the procedure (from the manual). The outcome is predetermined by the teacher and may also be already known to the learner. Expository instruction has been criticized for placing little emphasis on thinking (*135, 136*). Its 'cookbook' nature emphasizes the mechanical following of specific procedures to collect data, in order to verify or demonstrate principles described in textbooks. In this way, it is an ineffective means of building concepts, while little meaningful learning may take place (*139*). Such laboratory experiences facilitate the development of lower-order cognitive skills, such as rote learning and algorithmic problem solving (*140*). In addition, they "have little relevance to real life and so fail to promote in students a genuine interest and motivation for practical work" (*137*).

Inquiry-based activities are more student-centered, contain less direction, and give the student more responsibility for determining procedural options than the traditional format. They effectively give students ownership of the laboratory activity, which can result in the students' showing improved attitudes towards laboratories. Although this type of practical work can foster many of the aims, it is time consuming, potentially costly, and very demanding on those who have to organize large laboratory classes. However, there is a strong case for its use from time to time and at all levels, as a short inquiry, attached to the end of an expository laboratory (137). The *Journal of Chemical Education* publishes frequently innovative laboratory experiments that are inquiry- as well as discovery- and problem-based (search under "Inquiry-Based/Discovery Method" and "Problem-Based Learning").

Accepting the facts that the conventional expository laboratory has many problems associated with it, as well as that it is not an easy task to replace it entirely with inquiry-type practical work, Tsaparlis and Gorezi (141) proposed a modification of a conventional, one-semester, expository physical chemistry laboratory to accommodate a project-based component. Eight project-type tasks were used, mostly taken from articles in *the Journal of Chemical Education,* which is a rich source. The conventional experiments remained intact in this approach, being simply enriched with the project-based component. Students working cooperatively carried out both the conventional and the project parts: in pairs for the conventional experiments, in groups of four for the project work.

The originality of the projects and the feeling of ownership and responsibility contributed to the dedication and enthusiasm of the students during the performance of the experiments. The writing of the report and the oral presentation of the project were the very important concluding parts of it, and at the same time very demanding. The evaluation of this work by the students showed that the majority of them were in favor of cooperative work. Project work was judged superior for the development of a number of abilities in students, with most important being communication skills. Abilities related to the psychology of learning also benefited. The connection of chemistry with everyday life, especially modern applications (such as lithium batteries, commercial soaps, and corrosion of metals) attracted the students' interest and attention.

Innovative Methods of Teaching

Context-Based Approaches

Context-based approaches to teaching and learning provide applications as starting points from which to develop the subject. The success of such approaches in high-school teaching in the US with ChemCom (142) and in UK with the Salters chemistry course (143) has encouraged the use of the approach in university. In the US, this success was attributed, at least in part, to higher levels of interest and motivation amongst the students, together with their perception of the relevance of the topics (*144*). Zielinski and Schwenz (*145*) have identified the importance of context-rich teaching materials in the teaching of physical chemistry.

Holman and Pilling (*146*) designed a short thermodynamics course for first-year undergraduates that dealt with gas laws and the first law of thermodynamics. In this, they 'infused' contexts into the existing course. Research-linked contexts were thought appropriate for a university-level course. Context was applied to the course in a number of ways (lecture snippets, contextualized tutorial problems, workshop problems, and assessment questions). Short demonstrations and video-clips were used in lectures as a different kind of context. The approach was tested on students at two British universities. An evaluation of the course (by means of an end-of-year questionnaire) showed that the new approach contributed to making thermodynamics interesting and its principles clearer. Encouraging in favor of the contextualized learning was also the data from performance in the relevant examination. However, some doubts were expressed over how successful it was in developing students' abilities in problem solving.

Yang et al. (*147*) used batteries and flashlights as a real-life context. They depicted how a flashlight works using a computer animation representing the internal system of the flashlight with batteries in it. The project identified students' misconceptions about electricity and electrochemistry, and was reported to be successful in achieving conceptual change in students.

A context-based approach has also been developed by Belt et al. (*148*) for teaching aspects of thermodynamics, kinetics and electrochemistry usually associated with the early stages of undergraduate chemistry courses. The context was that of the generation of energy for an emerging city. Working in groups, students used an array of physical chemistry principles to examine the combustion of fossil fuels and hydrogen, the use of hydrogen in fuel cells, solar power, and energy from a geothermal source. Students' answers to feedback questionnaires revealed that the majority of them welcomed the opportunity to "put theory into practice" by studying physical chemistry in an applied context and to work as part of a group.

Context-based approaches are consistent with Novak's theory of human constructivism (see also Discussion below).

Teaching with Technology

The use of technology, and especially of computers, in undergraduate physical chemistry may contribute to better teaching and learning. In particular, the use of models, simulations and animations may help students contradict and overcome common misconceptions.

A survey of computer use in university courses has been carried out by Miles and Francis (*149*). Due to poor response rate of a Web-based questionnaire, the findings were skewed toward those faculty members who were using computers in their course. The most widely used application among the respondents was the spreadsheet, while the use of molecular modeling, and symbolic mathematics software lagged behind. The authors recommended that teachers should develop more computer-based activities, exploit fully the Web resources, and transfer effectively technical skills. They also provided a list of relevant Web sites, but in the time elapsed since that publication, the currently available Web material must be greatly enhanced, both in terms of number of materials available and in terms of quality of content.

Brattan, et al. (*150*) used computer-based resources for supporting the first-year physical chemistry laboratory. A number of CD packages, each comprising video, background theory, worked examples and sample data, a glossary and a final test were created. Three learning packages, one each for basic phase equilibriums, basic chemical kinetics, and gas chromatography, were tested with students. It was reported that the students made good use of the material. The use of idealized computer-generated data had a clear benefit, demonstrating the effect of experimental error on results. Students' attitude to laboratory work improved. They worked consciously at the computer, discussed with their peers, and asked questions. In the laboratory, they appeared confident in their ability to plan and execute the experiments with minimum input from supervisors.

The *Journal of Chemical Education* (JCE) provides a full range of computerized instructional materials, including Web-based computations and animations (search under "Computer-Based Learning", "Teaching with Technology", and "JCE WebWare: Web-Based Learning Aids"). According to Zielinski (*151*), in chemistry, but especially in physical chemistry, students "need excellent tools to enhance the potential for continued learning." For instance, in the April 2004 issue of JCE, Zielinski describes five Mathcad templates that deal with pressure-volume work, the Boltzmann distribution, the Gibbs free-energy function, intermolecular potentials and the second virial coefficient, and quantum mechanical tunneling.

A new collection of the JCE Digital Library is *JCE LivTexts: Physical Chemistry* (*152*). Its aim is to include instructional resources that span the physical chemistry curriculum. Each chapter includes links to supplementary learning objects, such as background material, advanced treatments, interactive symbolic mathematics lessons, and exercises and projects. For instance, "Quantum states of atoms and molecules" is an introduction to quantum mechanics applied to spectroscopy, the electronic structure of atoms and molecules, and molecular properties (*152*). According to its authors, it allows students to develop information processing, critical thinking, analytical reasoning, and problem-solving skills.

Finally, Web-based simulations, incorporated as end-of-chapter problems, are now used in physical chemistry textbooks (*153*), with the aim that "students can focus on the science (the concepts) and avoid a math overload."

Student Active and Cooperative Learning Methods

The major conclusion of science education research is that active-learning methods of teaching and learning should replace the traditional didactic method of lecture. Active learning may be implemented by students working on their own (under the instructor's observation and guidance), but more effectively by students working together in small groups: of four or five to accomplish an assigned common learning task/goal (*154*). Active and cooperative learning methods are consistent with *social/cultural constructivism*, provide a better learning environment, and contribute to deeper understanding and development of learning skills (*155-157*). This form of learning is traditionally used in laboratory work. It was also used in the project-enriched physical chemistry laboratory described above (*141*).

With regard to the size of the groups in that work, working in groups of four seemed acceptable to the majority of students (*141*). There have been studies (*158*) that suggested that pairs function better because peers cannot withdraw and leave the responsibility of the discussion to others. On the other hand, larger groups (e.g. fours) allow students to consider a wider range of ideas (*159*). Additional research reported that students progressed significantly more in their physics reasoning when interacting in fours rather than in pairs (*160*).

A crucial issue in cooperative work is the *dynamics* within the group. Each individual engaged in such work may play three roles: the *learner*, the *learner facilitator*, and the *leader.* Important is also the interaction mechanism that focuses on a *synergetic effect* among individuals. It is reasonable to suggest that under certain conditions (cognitive and affective ones), there is a possibility that the outcome of working in groups could surpass individual capabilities because cross-fertilization occurs between interacting peers. In any case, the extent to which cooperative work promotes an individual's acquisition and retention of learned material depends on many factors, such as: the individual differences of the group members, the nature of the task, the process itself, and prior training in group-skills. The unequal contribution of the members of the group appears to be a very serious problem of cooperative work - it occurred with about one third of the students in the physical chemistry laboratory projects (*141*).

Towns and Grant (*161*) studied cooperative learning activities in physical chemistry, in particular in a graduate-level thermodynamics course. Their purpose was to describe the structure of events during these activities. The findings showed that students were moved away from rote learning strategies toward more meaningful learning ones, which allowed them to integrate concepts. The title of the relevant paper, "I believe I will go out of this class actually knowing something," speaks for itself. In addition, students developed interpersonal and communication skills. Two issues of concern should be taken into account (*161*). First, simply placing the students into groups does not mean that they will function as a group; they need to be carefully prepared for cooperative learning. Second, low-status students do not verbalize their ideas as often as high-status students. This has direct bearing on assessment. The authors recommended focusing assessment on how the students function as a group.

Hinde and Kovac (*162*) used two active learning methods of physical chemistry, and compared and contrasted their relative strengths and weaknesses. The two authors employed quite different strategies, leading to two extremes of active learning methods. One author taught physical chemistry for majors, supplementing the traditional lecture method with cooperative-learning sessions that explored selected sessions in more depth. The sessions were heavily based on the use of computers, and were held approximately every two weeks in the computer laboratory. The other author taught biophysical chemistry to mostly biochemistry and biology students, and used almost exclusively cooperative learning, supplementing it occasionally with mini-lectures. In the second case, the students used a preliminary version of a set of guided-inquiry material (which relied almost exclusively on pencil-and-paper calculations) (see also the POGIL Project, below). Students' responses to the active-learning methodology were positive, and, more importantly, the students became more positive about physical chemistry. Guided inquiry turned out to be a more difficult method for teaching physical chemistry: "It is very difficult for students to extract the "big picture" from guided-inquiry exercises". A proper balance between exposition and guided inquiry should exist, and this depends on the students. In particular, it was found that students in biophysical chemistry would have benefited more form lectures, while students in physical chemistry for majors would have been benefited from more active learning exercises. Process-oriented guided-inquiry learning (POGIL) is a recent development in the teaching of physical chemistry. For a detailed description of that method, please see the chapter by Spencer and Moog in this book.

Discussion and Recommendations for Instruction

As stated in the introduction, science learning involves a complex interplay between the *process of learning* and the *content*. Knowledge of the content is a necessary but not sufficient condition for teaching chemistry (or any other subject); it is knowledge of the process of learning and the learner that provides the sufficient condition. Content knowledge refers to one's understanding of the subject matter, while pedagogical knowledge refers to one's understanding of teaching and learning processes independent of subject matter. The combination of content knowledge with pedagogical knowledge leads to *pedagogical content knowledge* (PCK) (*163, 164*). PCK refers to knowledge about the teaching and learning of particular subject matter that takes into account the particular learning demands inherent in the subject matter. According to Geddis (*165*), many of the pedagogical skills of the outstanding teacher are content-specific.

In this chapter, we saw studies that examined the factors that are most important to success in physical chemistry (basic mathematical skills, logical thinking skills, motivation, and study skills). In addition, we reviewed findings of research on students' misconceptions, and the difficulties in understanding deeply the concepts of physical chemistry, and particularly in the areas of thermodynamics, electrochemistry, and quantum chemistry. It was also suggested that problem solving is a very complicated process, involving more than one cognitive variable, as well as affective ones. A number of distinct skills, such as mathematical ability, information-processing and disembedding ability (but also scientific reasoning, and logical thinking skills) contribute to novel problem solving. Mathematical ability is necessary but not sufficient for success in problem solving. Providing students practice in using multiple representations, especially symbolic/pictorial, may contribute to improvement of problem-solving ability. Models are crucial in physical chemistry education. Research has shown that students prefer simple and realistic models, and find it difficult to replace the earlier models; instead, they often construct alternative models that contain elements from all earlier models. Of high importance is laboratory instruction, with the conventional expository laboratory being criticized for placing little emphasis on thinking. Inquiry-based activities require the learners to generate their own procedures. A simpler variety is provided by the project-based activities that, without calling for new procedures, require the students to try a new modern experiment. Finally, innovative approaches (such as context-based teaching, the use of new technology and the Internet, and active and cooperative learning) hold promise to affect students both cognitively and attitudinally.

Research in science education has shown that students derive from their physical and social experiences numerous *alternative conceptions* or *misconceptions*. [These include *instructor-driven misconceptions* (*166*) that result from previous elementary, imprecise, incomplete instruction.] The literature uses an "inventory approach", organizing and reporting misconceptions by topic or subject, and this we did here for various areas of physical chemistry. An common explanatory framework about the behavior of the natural world has been proposed recently by Talanquer for analysing misconceptions (*167*). It is based on "commonsense reasoning" that tends "to generate quick explanations based on intuition and broad generalizations, without much reflection." A number of *empirical assumptions*, and of *reasoning heuristics* have been proposed as responsible for causing misconceptions. The assumptions are:

continuity, substantialism, essentialism, mechanical causality, and teleology. The heuristics are: association, reduction, fixation, and linear sequencing. For instance, the continuity assumption causes the misconception that molecules have the macroscopic properties of the corresponding substance, while the substantialism assumption that electron clouds and shells are made of some kind of stuff; on the other hand, the association heuristic leads to the misconception that heat is the principal cause of all chemical reactions, while the linear-sequencing heuristic that in an electrochemical cell, electrons travel around the circuit causing a linear sequence of events.

Science education research on concept learning has shown that to change misconceptions is a very hard task. Traditional instruction cannot overcome these ideas. But some innovative strategies from science education research on *conceptual change* are promising. Hewson and Hewson (*168*) describe teaching strategies, which aim at linking ideas. One technique is *integration*, which attempts to link concepts, for example atomic orbitals and molecular orbitals. Another technique is *differentiation,* which tries to identify differences between related concepts, such as complex and real, or hydrogenic and non-hydrogenic orbitals. These two techniques can provide missing links among concepts and thus facilitate learning and overcome misconceptions.

Although it is recognized that most concepts of physical chemistry are highly abstract and complicated, we must also take note of a major source of difficulty in dealing with scientific concepts. It is known that many chemistry instructors subscribe to the view that mastery of chemical skills, such as manipulating mathematical equations and algorithmic problem solving, presupposes and therefore is equivalent to, conceptual understanding. Research reviewed above has shown that the ability, for instance, of students to write down electron configurations for atoms does not guarantee understanding of the underlying concepts. On the other hand, the need to cover a lot of material in an ever-expanding subject such as physical chemistry may lead instructors to rush through the material. We need to focus on *meaningful learning*, discourage rote learning, and aim at *coherent understanding*, and anticipate pre-conceptions. These can be achieved only by integrated, in-depth coverage of the topics (*168*).

The content of physical chemistry is crucial for teaching and learning. Zielinski and Schwenz (*169*) have addressed what content should be included in lecture and laboratory, and have connected this with the suggestions from educational research about how students learn and the teaching pedagogy. The authors paid particular attention to the way mathematics should be treated and used in physical chemistry as well as to differentiating content and emphasis according to the diverse needs of various future scientists. Enrichment of core concepts with modern examples from the literature was also emphasized. Relevant to this is the criticism by Scerri (*18*) of the way chemistry education research engages issues of content. Accordingly, more effort should be put "into trying to bridge the widening gap between front-line research in chemistry and the curriculum." Project cooperative work both in classes and labs could contribute here. Also, the content should not be left "as a mysterious black-box that is supposed to look after itself," as various innovative visualization multimedia projects tend to imply despite their ingenuity (*18*).

What is equally or even more important is that the overall teaching methodology must change, for it could be that more and better content, taught in the old didactic way, is unlikely to improve the situation. Students must be more actively and cooperatively involved in the construction of physicochemical knowledge. Conceptual-change, constructivist pedagogy holds promise of being more effective (*170*). According to Niaz (*171*), conceptual change requires that we include in our courses "the experimental details not as a 'rhetoric of conclusions', but as 'heuristic principles', that are based on arguments, controversies, and interpretations of the scientists."

Learning theories can guide physical chemists into expanding the traditional lecture by adding other activities. According to Novak's theory of human constructivism (*24, 25*) "meaningful learning will occur when education provides experiences that require students to connect knowledge across the cognitive, the affective, and the psychomotor domains. Given that the learner's and teachers' 'worlds' are not 'isomorphic' with one another, education must provide experiences for the sharing and negotiation of meanings. It should consist of those experiences that will empower a person to manage his/her daily life." Bretz (*25*) provided a very useful example of applying Novak's theory to physical chemistry teaching: students who study energy transformations must not only think of concepts (e.g. enthalpy and energy), but must also design and carry out experiments in the laboratory (e.g. combustion of foods with varying fat contents) which allow them to connect these abstract concepts to choices they must make in their daily lives (feeling and empowerment). Learning experiences that lack one or two of the three domains will prevent students from succeeding at meaningful learning.

Towns (*172*) has applied Kolb's experiential learning theory (*173*) to teaching physical chemistry (atomic structure and spectra, and rotational-vibrational spectroscopy). Kolb maintains that the learning is a continuous process that is grounded on experience. Therefore, learning is not the same for all individuals. There are two different processes of perceiving experience: sensing/feeling (for concrete experiences) and abstract conceptualisation (for thinking). In addition, there are two opposing ways of transforming experience: reflective observation (watching) and active experimentation (doing). In this way, individuals are categorized into *accommodators* (concrete and active), *divergers* (concrete and reflective), *convergers* (abstract and active), and *assimilators* (abstract and reflective) (*172*).

Moore and Schwenz (*37*) called on instructors to present the material of physical chemistry in a manner that excites students, illustrates the usefulness of the material, and generates an understanding of the chemistry, rather than a series of mathematical abstractions upon which the foundations of chemistry are laid. According to Engel and Reid (*153*), mathematics, while being central to physical chemistry can distract students from "seeing" the underlying concepts. Carson and Watson (*63*) suggested that undergraduate education in chemistry should be modified so that the majority of conceptual problems are addressed throughout the curriculum. Sözbilir (*52*) recommended that qualitative understanding of the concepts and laws of physical chemistry should precede the mathematical derivations and numerical calculations. Ogude and Bradley (*73*) maintained that traditional methods of teaching that pay attention to quantitative aspects of the subject are inadequate for detecting and overcoming misconceptions, while qualitative interpretations, that are mostly ignored by teachers, should be paid attention to.

We conclude with reference to the declaration of eleven US academics and scientists, published in *Science* magazine (*174*), that the time has come for '*scientific teaching*' to be used in universities, "in which teaching is approached with the same rigor as science at its best." "Scientific teaching involves active learning strategies to engage students in the process of science and teaching methods that have been systematically tested and shown to reach diverse students."

Acknowledgment

The author expresses his gratitude to Prof. George M. Bodner who suggested the writing of this review. He is also indebted to Dr. Stephen Breuer who read an earlier version of this manuscript and made suggestions for improvements. Finally, thanks are due to the two anonymous reviewers who contributed with their comments and suggestions to the substantial improvement of this chapter.

References

1. Bunce, D.; Robinson, W. R. *J. Chem. Educ.,* **1997**, *74*, 1076-1079.
2. Herron, J. D.; Nurrenbern, S. C. *J. Chem. Educ.,* **1999**, *76*, 1354-1361.
3. Task Force on Chemical Education Research. *J. Chem. Educ.,* **1994**, *71*, 850-852.
4. Caliendo, S. M.; Keele, W. C. *J. Res. Sci. Teach.,* **1996**, *33*, 225-227.
5. Johnstone, A. H. *Chem. Educ. Res. Pract.,* **2000**, *1*, 9-15.
6. Bodner, G. *J. Chem. Educ.,* **2001**, *78*, 1107.
7. Herron J. D. *J. Chem. Educ.,* **1975**, *52*, 146.
8. Herron, J. D. *J. Chem. Educ.,* **1978**, *55*, 165-170.

9. Bodner, G. M. *J. Chem. Educ.*, **1986**, *63*, 873-878.
10. Nurenburn, S.C. *J. Chem. Educ.*, **2001**, *78*, 1107-1110.
11. Shayer, M.; Adey, P. *Toward a Science of Science Teaching;* Heinmann Educational Books: London, 1981.
12. Bunce, D.M. *J. Chem. Educ.*, **2001**, *78*, 1107.
13. Driver, R. *The Pupil as Scientist?* Open University Press: Milton Keynes, 1983.
14. Taber, K. S. *Chemical Misconceptions – Prevention, Diagnosis and Cure*; Royal Society of Chemistry: London, 2002; Vol. 1.
15. Nola, R. *Sci. & Educ.*, **1997**, *6*, 55-83.
16. Niaz, M.; Abd-El-Khalick, F.; Benarroch, A.; Cardellini, L.; Laburú, C. E.; Marín, N.; Montes, L. A.; Nola, R.; Orlik, Y.; Scharmann, L. C.; Tsai, C.-C.; Tsaparlis, G. *Sci. & Educ.*, **2003**, *12*, 787-797.
17. Gil-Pérez, D.; Guisásola, J.; Morena, A.; Cachapuz, A.; Pessoa de Carvalho, A. M.; Martínez Torregrosa, J.; Salinas, J.; Valdés, P.; González, E.; Gené Duch, A.; Dumas-Carré, A.; Tricárico, H.; Gallego, R. *Sci. & Educ.*, **2002**, *11*, 557-571.
18. Scerri, E. R. *J. Chem. Educ.*, **2003**, *80*, 468-477.
19. Matthews, M. R. *Sci. & Educ.*, **2000**, *9*, 491-505.
20. Vygotsky, L. *Thought and Language;* MIT Press: Cambridge, MA, 1962.
21. von Glasersfeld, E. *Synthese*, **1989**, *80*, 121-140.
22. von Glasersfeld, E. *Radical Constructivim: A Way of Knowing and Learning*; Falmer Press: London, 1995.
23. Kelly, G. A. *The Psychology of Personal Constructs: A Theory of Personality*; W. Norton & Co.: New York, 1955.
24. Novak, J. D. *Learning, Creating, and Using Knowledge*; Lawrence Erlbaum: Mahwah, 1998.
25. Bretz, S. L. *J. Chem. Educ.*, **2001**, *78*, 1107.
26. Eylon, B.-S.; Linn, M. C. *Rev. Educ. Res.*, **1988**, *58*, 251-301.
27. Strike, K. A.; Posner, G. J. In *Cognitive Structure and Conceptual Change*; West, L.; Pines, A., Eds.; Academic Press, 1985.
28. Tsaparlis, *J. Chem. Educ.*, **1997**, *74*, 922-925.
29. Adey, P. *Sci. Educ.*, **1987**, *71*, 5-7.
30. Cohen, L.; Holliday, M. *Statistics for Social Sciences*; Harper and Row: London, 1982.
31. Kempa, R. K. *Assessment in Science*; Cambridge University Press: Cambridge, 1986; Chapter 3.
32. Marton, F. *Instructional Science*, **1981**, *10*, 177-200.
33. Niaz, M. *Sci. & Educ.*, **1997**, *6*, 291-300.
34. Tobin, K. NARST News, **1993**, *35*, 2.
35. Yeany, R. H. NARST News, **1992**, *34*, 1.
36. Pigot, I. In *Empirical Research in Chemistry and Physics Education;* Schmidt, J.-H., Ed.; ICASE: Dortmund, Germany, 1992; pp 113-134.

37. Moore R. J.; Schwenz, R. W. *J. Chem. Educ.,* **1992**, *69*, 1001-1002.
38. Ausubel, D. P. *The Acquisition and Retention of Knowledge: A Cognitive View*; Kluwer Academic Publishers: Dordrecht, the Netherlands, 2000; pp 40, 51.
39. Novak, J. D. *Sci. Educ.,* **2002**, *86*, 548-571.
40. Justi, R. S.; Gilbert, J. K. *Int. J. Sci. Educ.,* **2003**, 25, 1369-1386.
41. Grosslight, L.; Unger, C.; Jay, E.; Smith, C. *J. Res. Sci. Teach.,* **1991**, *29*, 799-822.
42. Gilbert, J. K.; Boulter, C. J.; Elmer, R. In Developing Models in Science Education; Kluwer Academic Publishers: Dordrecht, the Netherlads, 2000.
43. Harrison, A. G.; Treagust, D. F. *Sci. Educ.,* **1996**, *80*, 509-534.
44. Fishler, H.; Lichtfeldt, M. *Int. J. Sci. Educ.,* **1992**, *14*, 181.
45. Nicoll, G. *Int. J. Sci. Educ.,* **2001**, *23*, 707-730.
46. Petri, J.; Niedderer, H. *Int. J. Sci. Educ.,* **1998**, *20*, 1075-1088.
47. Coll, R. K.; Taylor, N. *Chem., Educ. Res. Pract.,* **2002**, *3*, 175-184
48. Vosniadou, S.; Brewer, W. F. *Cognitive Psychology,* **1992**, *24*, 535-585.
49. Nicoll, G.; Francisco, J. F. *J. Chem. Educ.,* **2001**, *78*, 99-102.
50. Derrick, M. E.; Derrick, F. W. *J. Chem. Educ.,* **2002**, *79*, 1013-1016.
51. Hahn, K. E.; Polik, W. F. *J. Chem. Educ.,* **2004**, *81*, 567-572.
52. Sözbilir, M. *J. Chem. Educ.,* **2004**, *81*, 573-578.
53. Herron, J, D. *The Chemistry Classroom: Formulas for Successful Classroom Teaching*; American Chemical Society: Washington, DC.; p 18.
54. Hudson, H. T.; McIntire, W. R. *Am. J. Phys.,* **1977**, *45*, 470-471.
55. Liberman, D.; Hudson, H. T. *Am. J. Phys.,* **1979**, *47*, 784-786.
56. Hudson, H. T.; Liberman, D. *Am. J. Phys.,* **1982**, *50*, 1117-1119.
57. Griffith, W. T. *Am. J. Phys.,* **1985**, *53*, 839-842.
58. Tiberghien, A., In *Proceedings of the First International Workshop Research on Physics Education*; Editions du CNRS: Paris, 1983; pp 73-90.
59. Brook, A.; Briggs, H.; Bell, B.; Driver, R. *Aspects of Secondary Students' Understanding of Heat – Full Report*. Centre for Studies in Science and Mathematics Education, University of Leeds: Leeds, 1984.
60. Kesidou, S. ; Duit, R. *J. Res. Sci. Teach.,* **1993**, *30*, 85-106.
61. Banerjee, A. C. *J. Chem. Educ.,* **1995**, *72*, 879-881.
62. Thomas, P. L.; Schwenz, R. W. *J. Res. Sci. Teach.,* **1998**, *35*, 1151-1160.
63. Carson, E. M.; Watson, J. R. *Univ. Chem. Educ.,* **1999**, *3*, 46-51.
64. Carson, E. M.; Watson, J. R. *Univ. Chem. Educ.,* **2002**, *6*, 4-12.
65. Sözbilir, M. *Univ. Chem. Educ.,* **2002**, *6*, 73-83.
66. Teichert, M. A.; Stacy, A. M. *J. Res. Sci. Teach.,* **2002**, *39*, 464-496.
67. Tarsitant, C.; Vicentini, M. *Sci. & Educ.,* **1996**, *5*, 51-68.
68. De Berg, K. C. *Sci. & Educ.,* **2006**, *15*, 495-519.
69. De Jong, O. ; Acampo, J.; Verdonk. A. *J. Res. Sci. Teach,* **1995**, *32*, 1097-1110.
70. Garnett, P. J.; Treagust, D. F. *J. Res. Sci. Teach.,* **1992**, *29*, 121-142.

71. Garnett, P. J.; Treagust, D. F. *J. Res. Sci. Teach.*, **1992**, *29*, 1079-1099.
72. Birss V. I.; Truax D. R. *J. Chem. Educ.*, **1990**, *67*, 403-409.
73. Ogude, A. N.; Bradley, J. D. *J. Chem. Educ.*, **1994**, *71*, 29-34.
74. Sanger, M. J.; Greenbowe, T. J. *J. Res. Sci. Teach.*, **1997**, *34*, 377-398.
75. Sanger, M. J.; Greenbowe, T. J. *J. Chem. Educ.*, **1999**, *76*, 853-860.
76. Özkaya A. R. *J. Chem. Educ.*, **2002**, *79*, 735-738.
77. Özkaya, A. R.; Uce, M.; Şahin, M. *Univ. Chem. Educ.*, **2003**, *7*, 1-12.
78. Niaz, M. *Int. J. Sci. Educ.*, **2002**, *24*, 425-439.
79. Greenbowe, T. J. *J. Chem. Educ.*, **1994**, *71*, 555-557.
80. Sanger, M. J.; Greenbowe, T. J. *J. Chem. Educ.*, **1997**, *74*, 819-823.
81. Harrison, A. G.; Treagust, D. F. *Sci. Educ.*, **2000**, *84*, 352-381.
82. Taber, K. S. *Chem. Educ. Res. Pract.*, **2002**, *3*, 145-158.
83. Taber, K. S. *Chem. Educ. Res. Pract.*, **2002**, *3*, 159-173.
84. Taber K.S. *Sci. Educ.*, **2005**, *89*, 94-116.
85. Tsaparlis, G.; Papaphotis, G. *Chem. Educ. Res. Pract.*, **2002**, *3*, 129-144.
86. Zoller, U. *J. Res. Sci. Teach.*, **1990**, *27*, 1053-1065.
87. Nakiboglu, C. *Chem. Educ. Res. Pract.*, **2003**, *4*, 171-188.
88. Coll, R. K.; Treagust, D. F. *Res. Sci. Educ.* **2001**, *31*, 357-382.
89. Coll, R. K.; Treagust, D. F. *Res. Sci. Tech. Educ.*, **2002**, *20*, 241-267.
90. Kalkanis, G., Hadzidaki, P.; Stavrou, D. *Sci. Educ.*, **2003**, *87*, 257-280.
91. Greca, I. M.; Freire Jr., O. *Sci. & Educ.*, **2003**, *12*, 541-557.
92. Shiland, T. W. *Int. J. Sci. Educ.*, **1997**, *34*, 535-545.
93. Tsaparlis, G. *Res. Sci. Educ.*, **1997**, *27*, 271-287.
94. Pauling, L.; Wilson, E. B. Jr. *Introduction to Quantum Mechanics with Applications to Chemistry;* McGraw-Hill: New York, 1935; p iii.
95. Coulson, C. A. *Chemistry in Britain,* **1974**, *10*, 16-18.
96. Simons, J. J. *Chem. Educ.*, **1991**, *68*, 131-132.
97. Castro, E. A.; Fernandez, F. M. *Int. J. Sci. Educ.*, **1987**, *9*, 441-447.
98. Birkhoff, G.; von Newmann, J. *Annals Mathematics,* **1936**, *37*, 835-843.
99. Fong, P. *Elementary Quantum Mechanics;* Addison-Wesley: Reading MA, 1962; pp 45-52.
100. Karakostas, V.; Hadzidaki, P. *Sci. & Educ.*, **2005**, *14*, 607-629.
101. Tsaparlis, G. *Chem. Educ. Res. Pract.*, **2001**, *2*, 203-213.
102. Scerri, E. R. *Chem. Educ. Res. Pract.*, **2001**, *2*, 165-170.
103. Sánchez Gómez, P. J.; Martín, F. *Chem. Educ. Res. Pract.*, **2003**, *2*, 131-148.
104. Scerri, E. R. *J. Chem. Educ.*, **2000**, *77*, 522-525.
105. Scerri, E. R. *J. Chem. Educ.*, **1998**, *75*, 1384-1385.
106. Scerri, E. R. *J. Chem. Educ.*, **1999**, *76*, 608.
107. Johnstone, A. H. In *Creative Problem Solving in Chemistry;* Wood., C.; Sleet, R., Eds.; The Royal Society of Chemistry: London, 1993; pp iv-vi.
108. Ritchie, I. M.; Thislethwaite, P. J.; Craig, R. A. *Problems in Physical Chemistry;* Jon Wiley & Sons: Sydney, 1975.

109. Zoller, U. *J. Chem. Educ.*, **1993**, *70*, 195-197.
110. Zoller, U.; Tsaparlis, G. *Res. Sci. Educ.*, **1997**, *27*, 117-130.
111. Simon, D. P.; Simon, H. A. In *Childrens' Thinking: What Develops;* Siegler, R.S., Ed.; Lawrence Erlbraum: Hillsdale NJ, 1978.
112. Larkin, J. H. In *Problem Solving and Education: Issues in Teaching and Research*; Tuma, D. T.; F. Reif, F., Eds.; Lawrence Erlbaum Associates: Hillsdale NJ, 1980.
113. Reif, F. *Phys. Teacher,* **1981**, *19*, 329-363.
114. Larkin, J.; McDermott, J.; Simon, D.; Simon, H. A., *Science*, **1980**, *208*, 1335-1342.
115. Bodner, G.M.; McMillen, T. L. *J. Res. Sci. Teach.*, **1986**, *23*, 727-737.
116. Pribyl, J. R.; Bodner, G. M. *J. Res. Sci. Teach.*, **1987**, *24*, 229-240.
117. Tsaparlis G. *Res. Sci. Tech. Educ.*, **2005**, *23*, 125-148.
118. Westbrook, S. L.; Rogers, L. N. *J. Res. Sci. Teach.*, **1994**, *3*, 65-76.
119. Atkinson, R. C.; Shiffrin, R. M. In *The Psychology of Learning and Motivation: Advances in Research and Theory*; Spence, K.W., Ed.; Academic Press: New York, 1968; Vol. 2, pp 89-195.
120. Baddeley, A. D. *Working Memory*; Oxford University Press: Oxford, 1986.
121. Baddeley, A. D. *Human Memory: Theory and Practice*; Erlbaum: London, 1990.
122. Pascual-Leone, J. *Acta Psychologica,* **1970**, *32*, 301-345.
123. Pascual-Leone, J.; Goodman, D.; Ammon, P.; Subelman, I. In *Knowledge and Development*; Gallagher, J.M.; Easley, J.A., Eds.; New York, Plenum: New York, 1978; Vol. 2, pp 234-289.
124. Pascual-Leone, J. *Int. J. Psych.*, **1987**, *22*, 531-570.
125. Witkin, H. A.; Dyk, R. B.; Paterson, H. F.; Goodenough, D. R; Karp, S. A. *Psychological Differentiation - Studies of Development*; Wiley: New York, 1974.
126. Pascual-Leone, J. In *Cognitive Style and Cognitive Development;* Globerson, T.; Zelniker, T. Eds.; Ablex: Norwood, NJ, 1989; pp 36-70.
127. Demerouti, M.; Kousathana, M.; Tsaparlis, G. *Chem. Educator,* **2004**, *9*, 132-137.
128. Johnstone, A. H.; Al-Naeme, F. F. *Int. J. Sci. Educ.*, **1991**, *13*, 187-192.
129. Niaz, M. *Educ. Psychology,* **1994**, *14*, 283-290.
130. Stamovlasis, D.; Kousathana, M.; Angelopoulos, V.; Tsaparlis, G; Niaz, M. *Perceptual and Motor Skills,* **2002**, *95*, 914-924.
131. Tinajero, C.; Paramo, M. F. *Brit. J. Educ. Psychology*, **1997**, *67*, 199-212.
132. Bodner G.M.; Domin D.S., *Univ. Chem. Educ.*, **2000**, *4*, 24-30.
133. Tsaparlis, G.; Zoller, U. *Univ. Chem. Educ.*, **2003**, *7*, 50-57.
134. Frazer, M. J. *Chem. Soc. Rev.*, **1982**, *11*, 171-190.
135. Hofstein, A.; Lunetta, V. *Rev. Educ. Res.*, **1982**, *52*, 201-217.
136. Hofstein, A.; Lunetta, V. *Sci. Educ.*, **2004**, *88*, 25-54.

137. Johnstone, A. H.; Al-Shuaili, A. *Univ. Chem. Educ.*, **2001**, *5*, 42-51.
138. Domin, D. S. *J. Chem. Educ.*, **1999**, *76*, 109-112.
139. Johnstone, A. H.; Wham, A. J. B. *Educ. Chem.*, **1982**, *19(3)*, 71-73.
140. Meester, M. A. M.; Maskill, R. *Second Year Practical Classes in Undergraduate Chemistry Courses in England and Wales*; Royal Society of Chemistry: London, 1994.
141. Tsaparlis G.; Gorezi M., *Can. J. Sci. Math. Tech. Educ.*, **2005**, *5*, 111-131.
142. Sutman F. X.; Bruce M. H., *J. Chem. Educ.*, **1992**, *69*, 564-567.
143. Burton W. G.; Holman J.S.; Pilling G.M.; Waddington D.J. *J. Chem. Educ.*, **1995**, *72*, 227-230.
144. Gutwill-Wise J. P. *J. Chem. Educ.*, **2001**, *78*, 684-690.
145. Zielinski T. J.; Schwenz R. W. *J. Chem. Educ.*, **2001**, *78*, 1173-1174.
146. Holman, J.; G. Pilling, G. *J. Chem. Educ.*, **2004**, *81*, 373-375.
147. Yang, E.-M.; Greenbowe, T. J.; Andre, T. *J. Chem. Educ.*, **2004**, *81*, 587-595.
148. Belt, S. T.; Leisvik, M. J.; Hyde, A. J.; Overton, T. L. *Chem. Educ. Res. Pract.*, **2005**, *6*, 166-179.
149. Miles Jr., D. G.; Francis, T. A. *J. Chem. Educ.*, **2002**, *79*, 1477-1479.
150. Brattan, D,; Mason, D.; Best, A. J. *Univ. Chem. Educ.*, **1999**, *3*, 59-63.
151. Zielinski, T. J. *J. Chem. Educ.*, **2003**, *80*, 580-581.
152. Zielinski, T. J. *J. Chem. Educ.*, **2005**, *82*, 1880.
153. Engel, T.; Reid, P. *Physical Chemistry*; Pearson Education / Benjamin Cunnings: San Francisco, 2006; p v.
154. Flynn, A. E.; Klein, J. P. *Educ. Tech. Res. Development*, **2001**, *49(3)*, 71-84.
155. Johnson, D.W.; Johnson, R. T.; Smith, K. A. *Active Learning Cooperation in the Learning Classroom*; Interaction Book Co.: Edina, MN, 1991.
156. Duncan-Hewitt, W.; Mount, D. L.; Apple, D. A. *A Handbook on Cooperative Learning*, 2nd ed.; Pacific Crest: Corvallis, OR, 1995.
157. Nurrenbern, S. C.; Robinson, W. R. *J. Chem. Educ.*, **1997**, *74*, 623-624.
158. Webb, N. M. *Int. J. Educ. Res.*, **1989**, *13*, 21-39.
159. Needham, R. In *Children's Learning in Science Project*; Centre for Studies in Science and Mathematics Education, University of Leeds: Leeds, 1987.
160. Alexopoulou, E.; Driver, R. *J. Res. Sci. Teach.*, **1996**, *33*, 1099-1114.
161. Towns, M. H.; Grant, E.R. *J. Res. Sci. Teach.*, **1997**, *34*, 819-835.
162. Hinde, R. J.; Kovac, J. *J. Chem. Educ.*, **2001**, *78*, 93-99.
163. Shulman, L. S. *Educ. Researcher*, **1986**, *15*, 4-14.
164. Bucat, B. R. *Chem. Educ. Res. Pract.*, **2004**, *3*, 215-228.
165. Geddis, A. N. *Int. J. Sci. Educ.*, **1993**, *15*, 673-683.
166. Bodner, G. *J. Chem. Educ.*, **1991**, *68*, 385-388.
167. Talanquer, V. *J. Chem. Educ.*, **2006**, *83*, 811-816.
168. Hewson, W. H.; Hewson, M. G. A. *Instructional Science*, **1984**, *13*, 1-13.
169. Zielinski, T. J.; Schwenz, R. W. *Chem. Educator*, **2004**, *9*, 108-121.

170. Stofflett, R. T.; Stoddart, T. *J. Res. Sci. Teach.*, **1994**, *31*, 31-51

171. Niaz, M. *Sci. Educ.*, **2002**, *86*, 505-525.

172. Towns, M. H. *J. Chem. Educ.*, **2001**, *78*, 1107.

173. Kolb, D. A. *Experiential Learning: Experience as the Source of Learning and Development*; Prentice-Hall: Englewood Cliffs, NJ, 1984.

174. Handelsman, J.; Ebert-May, D.; Beichner, R.; Bruns, P.; Chang, A.; DeHaan, R.; Gentile, J.; Lauffer, S.; Stewart, J.; Tilghman, S. M.; Wood, W. B. **2004**, *Science, 304*, 521.

Laboratory Literature Review

Chapter 8

Modern Developments in the Physical Chemistry Laboratory

Samuel A. Abrash

Department of Chemistry, University of Richmond, Richmond, VA 23173

Developments in the physical chemistry laboratory since the publication of the germinal text by Schwenz and Moore (*1*) are categorized and reviewed. The categories examined include modern instrumentation, current topics in chemistry, integrated laboratories, and developments based on chemical education research. New experiments involving traditional instrumentation and topics are include but are not reviewed extensively.

Introduction

In 1993 Richard W. Schwenz and Robert J. Moore published a book, under the auspices of the American Chemical Society, entitled "Physical Chemistry: Developing a Dynamic Curriculum"(*1*). This book followed a 1988 project by the Pew Mid-Atlantic Cluster on revision of the physical chemistry laboratory curriculum, and NSF funded workshops in 1990 and 1991 on physical chemistry curriculum development. Together they called for substantial changes in the content of the physical chemistry lab.

The Schwenz and Moore book called for inclusion of modern laboratory instrumentation and techniques, as well as modern research topics in the laboratory curriculum. Under the umbrella of modern instrumentation, the authors included experiments with lasers, mass spectrometers and cyclic voltammetry. In modern topics, computational chemistry, experiments with biological relevance, atmospheric chemistry and polymer chemistry were included. In addition, an experiment on the physical aspects of NMR was included.

Interestingly, there were two significant gaps in the recommendations. The first was the lack of recommendations for education research on the physical chemistry laboratory, although in his contextual article (2), McCay called for future assessment of the experiments developed by the Pew Mid-Atlantic Cluster. The second related lacuna was in an absence of education research based revisions of the physical chemistry laboratory as a whole.

It has been 13 years since Schwenz and Moore was published. In this article I will be looking to review the developments in the physical chemistry laboratory since the publication of this germinal book. I am attempting to be reasonably comprehensive in this review, and to bring some order to the large numbers of published experiments, I will discuss them under a few broad headings, that, while based in part on the emphases in Schwenz and Moore, have been expanded to reflect the range of new work in the field. These include Modern Techniques, Experiments for Integrated Laboratories, Studies of Modern Materials, Education Research Based Experiments or Curricula, and experiments on topics chosen for their relevance, that I call Real World Experiments. In addition, there has been a great deal of activity by faculty who have developed or improved experiments using traditional instrumentation or devoted to more traditional topics. These are valuable as well, and I have devoted a section to these new experiments.

I have deliberately omitted those experiments already published in the form of laboratory manuals. One reason is the ready availability and clear organization of these textbooks. A second is that even for those with recent editions, most of the experiments they include predate Schwenz and Moore. Before embarking on this extensive review, however, I want to make a serious call for a major expansion in the effort and funding for education research on the physical chemistry laboratory. The experiments that I'm about to discuss have been the result of substantial thought and effort on the part of many of our colleagues, but with a very few exceptions, there has been no clear identification of the pedagogical goals, or the pedagogical benefits, and even fewer attempts to assess the degree to which they succeed. Some examples of reasonable pedagogical goals would include reinforcement of lecture topics, or instruction in laboratory techniques. The laboratory can be used to introduce a topic beyond that covered in the lecture, or to provide a context for lecture material. It can be used to teach research methods, experimental design or sophisticated methods of data analysis. Of course this is just a selection of possible goals. A good experiment could have one or more of these as its goals, or have others not mentioned. However, it is critical that a faculty member writing about a new experiment clearly identify the goals of the experiment, because without clear identification of the goals, it is not possible to determine the degree of pedagogical success of the experiment.

Even when a development is research based, in most cases the research is based on studies of first-year college students or on the National Research Council volumes, <u>How People Learn: Brain, Mind, Experience and School</u> (*3*) and <u>How Students Learn: History, Mathematics and Science in the Classroom</u> (*4*), that contain education research on students from K-12. Aristotle once said that the unexamined life is not worth living. Should we perhaps say that the unexamined curriculum is not worth teaching? We should all bring chemical education research into our courses by becoming aware of the research that has been done, by applying it to our teaching, and by opening our departments to the practitioners of this critical subject.

Complete Revisions and Chemical Education Based Approaches

As noted above, there are not many new approaches to the physical chemistry laboratory that are based on Chemical Education research, but there are some that deserve mention. Articles that are based on these approaches are listed in Table I.

Chemical Education Research Related to the Physical Chemistry Lab

Malina et al. (*5*) have published a constructivist analysis of how instrument use affects student concept development. The goal was to identify those characteristics that were the most important influences on student construction of scientific understanding. They focused on CCD Spectroscopy, and based their work on multiple theoretical frameworks, including distributed cognition and the theory of affordances.

Weaver (*6*), Long et al. (*7*), Slocum et al. (*8*), and Sauder et al. (*9*) have developed laboratory learning aids based on chemical education research and have done careful assessments of these aids. Weaver's project has been the development of a series of DVDs with the collective title "Physical Chemistry in Practice". The series currently includes disks on AFM and Surfaced Enhanced Raman Spectroscopy, and will eventually consist of a series of 10 DVDs covering a wide range of topics in the physical chemistry lecture and lab. They are meant to supplement either the lecture or laboratory portion of the course. The goal was to present modern topics in physical chemistry in a way that places them in a relevant context. The disks include theoretical material and graphical illustrations of how the techniques work, but also include illustrations of applications. Students using these DVDs along with their lecture or lab showed an increase in understanding of the material, but cited the understanding of the reasons for these experiments arising from the discussions of applications as the most important feature of the DVDs.

Another project that aims, in part, to enhance laboratory experiments is the Physical Chemistry Online (PCOL) project, that has been under development since 1994 by a large group of collaborators (7, 8, 9). PCOL consists of a series of short exercises involving faculty and students from several geographically distributed institutions working collaboratively. One goal of the project is to provide the opportunity for pedagogical innovation to faculty in departments that are sufficiently small that equipment, institutional or collegial support necessary for such innovation are not present (9). Several modules have been developed, on topics ranging from flame thermodynamics to laser spectroscopy. These modules use pedagogical approaches that include cooperative learning, case studies, and discovery-based learning. Rigorous assessment has been done to demonstrate the effectiveness of this project.

Deckert et al. (10) have based a complete revision of a physical chemistry laboratory course on chemical education research. The course consists of four three-week modules. In each of the modules, students are guided toward planning their own experiments and data analysis and work collaboratively. A goal is to increase student preparedness for independent research. Some assessment of this project has been done. Gourley (11) has also tried a wholesale revision with the goal of creating a more "research-like experience". The unique feature of this laboratory curriculum is that after working on a module for a fixed period of time, each group presents a progress report, and then passes the project on to a new group to be continued. This project is still in an early phase and has not yet been formally assessed.

In addition to these systemic approaches, I found two examples of individual experiments that were revised with an eye to the chemical education research. One is Long and coworkers' revisiting of the Spectrum of I_2 (12). Long et al. have developed a set of activities, based on education research, to facilitate the learning of the concepts critical to understanding this experiment. The approach has been carefully assessed and showed positive results both in student attitudes and understanding.

Experiments Involving Modern Instrumentation

In the past 13 years there has been a great deal of activity in the development of experiments using modern instrumental methods. While the bulk of these have been experiments either involving lasers, or magnetic resonance spectrometers, a very wide range of instruments and techniques are represented, including mass spectrometers, x-ray diffractometers, capillary electrophoresis, and the various molecular resolution electron spectroscopies. Because of the large numbers of experiments in each category, laser experiments

Table I. New Developments Based on Education Research

Title	Ref. #
How Students Use Scientific Instruments To Create Understanding: CCD Spectrophotometers	5
The Physical Chemistry in Practice DVD Series	6
Physical Chemistry Online: A Small-Scale Intercollegiate Interactive Learning Experience	7
Online Chemistry Modules: Interaction and Effective Faculty Facilitation	8
Physical Chemistry Online: Maximizing Your Potential	9
An Example of a Guided Inquiry, Collaborative Physical Chemistry Laboratory Course	10
Theory and Experiment	11
The Iodine Spectrum: A New Look at an Old Topic	12
Projects in the Physical Chemistry Laboratory: Letting the Students Choose	13

and magnetic resonance experiments will be treated in their own section. The remainder of the modern instruments and techniques will be treated together.

Experiments Using Lasers

The largest number of laser experiments were some form of fluorescence experiment (*14-25*). They are listed, along with transient absorption experiments (*26-28*), in Table II. In a way the large number of fluorescence experiments is not particularly surprising because fluorescence, as a zero-background technique, can be studied with relatively low-powered lasers and is therefore relatively inexpensive to implement. In fact, one of the experiments uses the ubiquitous and inexpensive Helium-neon laser to excite fluorescence in molecular iodine (*14*) while another illustrates several experiments that can be carried out using inexpensive diode lasers (*15*). Seven of these experiments used the relatively inexpensive pulsed-nitrogen laser (*16-22*). Among these laser fluorescence experiments are studies of excimer dynamics (*16, 17*), the use of luminescence quenching studies to determine pK_a (*18*), luminescence studies of porous silicon (*19*), and the determination of viscosity by studying fluorescence depolarization (*23*).

Only three experiments involving transient absorption were found, probably because the populations required for a good signal to noise ratio require more powerful and therefore more expensive lasers. One of these used transient absorption spectroscopy to study ozone formation (*26*), making this an experiment of environmental interest as well, the second studied the displacement of benzene from an inorganic complex (27), while the third demonstrated an inexpensive apparatus that could be used for a number of different experiments (28).

Another extremely well-represented group of experiments were those exploiting the Raman effect (29-35). Interestingly, in addition to the four experiments in this group involving simple Raman scattering (29-32), three involved nonlinear Raman effects. These included stimulated Raman scattering (33, 34), and surface enhanced Raman (35). These Raman experiments can be found listed in Table III.

Raman was not the only type of laser experiment based on scattering. Several experiments were developed that exploited detection and analysis of scattered light. These included the use of light scattering to study critical phenomena (36), to probe surfaces (37), and to monitor growth of nanoparticles (38).

Another set of experiments that exploited the coherence of laser emission used interferometry or refractometry to study a range of phenomena. These included polymerization kinetics (39), diffusion (40), surface tension (41), and viscosity (42). These experiments, along with the laser light scattering experiments, can be found in Table III.

A fascinating category of experiments can be found in Table IV. These are the use of lasers to determine thermodynamic parameters. These include calorimetry (43), enthalpies of vaporization and vaporization rates (44, 45), and heat capacities (46). Other laser experiments that can be found in Table IV include the use of CW laser spectroscopy to determine the iodine binding-energy curve (47), the study of vibrational line profiles to determine intermolecular interactions (48), two photon ionization spectrometry (49), a study of optical activity and optical rotatory dispersion (50) and the development of several experiments using blue diode lasers (51).

Magnetic Resonance Experiments

Some readers may wonder at the inclusion of NMR under the rubric of modern instrumentation. After all, NMR spectroscopy has been a part of the curriculum in Organic Chemistry for years, and it is a rare student who cannot use ^1H NMR, ^{13}C NMR, and a host of multiple-pulse techniques to identify even structures of moderate complexity. However, NMR as a technique goes far beyond structure determination, and a number of these facets have been included in recent physical chemistry experiments.

One of these applications of NMR is in the non-invasive determination of equilibrium constants. These are usually most easily carried out for unimolecular

Table II. Laser Experiments – Fluorescence and Transient Absorption

Title	Ref. #
The Helium-Neon Laser-Induced Fluorescence Spectrum of Molecular Iodine	14
Using a Diode Laser for Laser-Induced Fluorescence	15
Nanosecond Time-Resolved Fluorescence Spectroscopy in the Physical Chemistry Laboratory: Formation of Pyrene Excimer in Solution	16
The Nitrogen-Laser Excited Luminescence of Pyrene	17
Determinations of pK_a from Luminescence Quenching Data	18
Understanding the Origin of Luminescence in Porous Silicon	19
Halide (Cl^-) Quenching of Quinine Sulfate Fluorescence: A Time-Resolved Fluorescence Experiment for Physical Chemistry	20
Excited State Lifetimes and Bimolecular Quenching of Iodine Vapor	21
Laser-Induced Fluorescence of Lightsticks	22
Viscosity by Fluorescence Depolarization of Probe Molecules	23
Characterizing the Behavior and Properties of an Excited Electronic State: Electron-Transfer Mediated Quenching of Fluorescence	24
Fluorescence Lifetime and Quenching of Iodine Vapor	25
Introduction of Laser Photolysis-Transient Spectroscopy in an Undergraduate Physical Chemistry Laboratory: Kinetics of Ozone Formation	26
Displacement of the Benzene Solvent Molecule from $Cr(CO)_5$(benzene) by Piperidine: A Laser Flash Photolysis Experiment	27
Luminescence Decay and Flash Photolysis Experiments Using and Inexpensive, Laser-Based Apparatus	28

reactions (52, 53, 54), or for formation of complexed, either hydrogen bonded (54, 55) or some sort of chelation process (56). NMR is ideal for these types of experiments because it is it allows determination of equilibrium constants without quenching the reaction, and with minimum perturbation. Similar considerations allow NMR to be used to follow the kinetics of some systems, such as the hydrolysis of orthoesters (57), or the more novel study of the DCl-HBr isotope exchange reaction in the gas phase (58).

These experiments while moving beyond simple structure determination, still depend primarily on structure determination as their primary tool. A number of experiments have moved beyond this to exploit other characteristics of NMR. For example, spin saturation transfer can be used to study internal rotation of molecules (59, 60), and line shape analyses can be used to study intramolecular exchange processes (61), conformational interchange (62) and

Table III. Laser Experiments: Raman, Light Scattering and Interferometry

Title	Ref. #
A Modular Raman Spectroscopy Using a Helium-Neon Laser That Is Also Suited for Emission Spectrophotometry Experiments	29
Quantitative Determination of the Rotameric Energy Differences of 1,2-Dihaloethanes Using Raman Spectroscopy	30
Raman Spectroscopy of Symmetric Oxyanions	31
The Second-Order Raman Spectrum of ^{13}C Diamond: An Introduction to Vibrational Spectroscopy of the Solid State	32
A Nonlinear Optical Experiment: Stimulated Raman Scattering in Benzene and Deuterated Benzene	33
Stimulated Raman Spectroscopy of Small Molecules	34
Surface Enhanced Raman Spectroscopy: A Novel Physical Chemistry Experiment for the Undergraduate Laboratory	35
Binary-Solution Critical Opalescence	36
Surface Light Scattering Adapted to the Advanced Undergraduate Laboratory	37
Monitoring Particle Growth: Light Scattering Using Red and Violet Diode Lasers	38
An Interferometric Study of Epoxy Polymerization Kinetics	39
Diffusion of CsCl in Aqueous Glycerol Measured by Laser Refraction	40
Surface Tension Determination through Capillary Rise and Laser Diffraction Patterns	41
A Diode-Laser-Based Automated Timing Interface for Rapid Measurement of Liquid Viscosity	42

Table IV. Laser Experiments: Thermodynamic Measurements and Miscellaneous Topics

Title	Ref. #
Photoacoustic Calorimetry: An Undergraduate Physical-Organic Experiment	43
Raman Spectroscopic Determination of Heats of Vaporization of Pure Liquids	44
Measurement of Evaporation Rates of Organic Liquids by Optical Interference	45
Determination of Heat Capacity of Liquids with Time-Resolved Thermal Lens Calorimetry	46
Binding Energy Curve for Iodine Using Laser Spectroscopy	47
Vibrational Line Profiles As a Proof of Molecular Interactions	48
Two-Photon Ionization Spectrometry of Alkali Atoms in Flames	49
Using Guided Inquiry to Study Optical Activity and Optical Rotatory Dispersion in a Cross-Disciplinary Chemistry Lab	50
Blue Diode Lasers: New Opportunity in Chemical Education	51

electron transfer processes (*63*). Some experiments focus on fundamental aspects of NMR such as factors affecting spin-lattice relaxation times (*64, 65, 66*).

One interesting experiment combines several aspects of NMR spectroscopy, including multiple-dimensional NMR, the physics of spin systems, and the ability to study molecular organization, in the study of the spectra of a phospholipid (*67*). Another interesting experiment uses the difference in nuclear spin splittings observed for two different isotopes of boron to determine its isotopic ratio (*68*). Finally, one experiment combines 2D-NMR with computational chemistry in order to obtain complete assignment of terpene spectra (*69*).

In contrast to the large number of NMR experiments, there are relatively few Electron Spin Resonance (ESR) experiments (a.k.a. Electron Paramagnetic Resonance, EPR) for the physical chemistry laboratory. This may be in part because of the cost of the apparatus, and the relative difficulty in interpretation compared to NMR. Despite these challenges, two recent experiments were developed to bring the ESR into the physical chemistry laboratory. One of these uses the technique to study the electronic structures of paramagnetic species (*70*). The other uses both an instructional and a research grade spectrometer to study the spectra of two inorganic complexes and one organic radical to introduce the students both to the principles of ESR, and the experience of spectral interpretation. (*71*). Both NMR and ESR experiments are listed in Table V.

Atomic Microscopy, Mass Spectrometry, X-Ray Diffraction, Electrochemistry and Miscellaneous Techniques

In addition to the techniques already discussed, experiments have been developed for the physical chemistry laboratory involving a number of other modern techniques. These are collected in Table VI. One group of techniques that has gotten a lot of recent attention in the literature involves atomic resolution microscopies, such as Scanning Tunneling Microscopy (STM) and Atomic Force Microscopy (AFM). One useful paper presents background to these techniques that would be a useful supplement for anyone teaching STM in lab (*72*). The labs involving these techniques include an investigation of intermolecular interactions (*73*), the use of STM to identify functional groups (*74*), a study of surface oxidation kinetics (*75*), and an introduction to AFM experiments (*76*).

Mass spectrometry was one of the techniques highlighted in Schwenz and Moore, but there has been relatively little activity in developing new experiments in this area. However, new experiments have been developed using GC/MS and GC/MS/MS to study ion fragmentation mechanisms (*77*), using ion-trap MS to study gas phase basicities of amino acids (78), and one series of experiments using a simple and relatively inexpensive quadrupole mass spectrometer (79).

X-ray diffraction has been a part of the physical chemistry laboratory curriculum for a long time, but mostly using the relatively simple powder diffraction technique. However a new experiment introduces the more complex method of single crystal X-ray diffraction (80). Another new experiment uses the technique to investigate the structure of alloys (81).

Table V. NMR and EPR Experiments

Title	Ref. #
The Cis-Trans Equilibrium of N-Acetyl-L-Proline	52
The Rearrangement of an Allylic Dithiocyanate	53
Hydrogen Bonding Using NMR: A New Look at the 2,4-Pentanedione Keto-Enol Tautomer Experiment	54
Hydrogen-Bonding Equilibrium in Phenol Analyzed by NMR Spectroscopy	55
The Complexation of the Na+ by 18-Crown-6 Studied via Nuclear Magnetic Resonance	56
Kinetics of the Hydrolysis of Orthoesters: A General Acid-Catalyzed Reaction	57
Gas Phase NMR Study of the DCl-HBr Isotope Exchange Reaction	58
NMR Determination of Internal Rotation Rates and Rotational Energy Barriers	59
Kinetics of Internal Rotation of N,N-Dimethylacetamide: A Spin-Saturation Transfer Experiment	60
Dynamic NMR of Intramolecular Exchange Processes in EDTA Complexes of Sc^{3+}, Y^{3+}, and La^{3+}	61
Conformation Interchange in Nuclear Magnetic Resonance Spectroscopy	62
Examination of Electron Transfer Self-Exchange Rates Using NMR Line-Broadening Techniques	63
Teaching the Fundamentals of Pulsed NMR Spectroscopy in an Undergraduate Physical Chemistry Laboratory	64
Determination of Spin-Lattice Relaxation Time Using ^{13}C NMR	65
Superoxygenated Water as an Experimental Sample for NMR Relaxometry	66
NMR of a Phospholipid	67
Measurement of the Isotopic Ratio of $^{10}B/^{11}B$ in $NaBH_4$ by ^1H NMR	68
Complete Assignment of Proton Chemical Shifts in Terpenes	69
Use of EPR Spectroscopy in Elucidating Electronic Structures of Paramagnetic Transition Metal Complexes	70
An EPR Experiment for the Undergraduate Physical Chemistry Laboratory	71

Schwenz and Moore introduced cyclic voltammetry as a modern approach to electrochemistry experiments. Three new experiments exploit this technique. One uses the technique as a probe or electrode surface area (*82*). A second uses the method to study adsorption of polyoxometalates on graphite electrodes (*83*). A third studies the effect of micelles on the diffusion and redox potentials of the well-studied ferrocene system (*84*).

In addition to these, there were experiments exploiting quartz crystal microbalances (*85*), matrix isolation (*86*), differential scanning calorimetry (*87*), capillary electrophoresis (*88*), and inductively coupled plasma (ICP) spectrometers (*89*).

Clearly the development of experiments involving modern instrumentation is a vibrant part of Physical Chemistry laboratory pedagogy. Equally clearly, there is now a need to focus more strongly on experiments involving instruments other than lasers or magnetic resonance spectrometers.

Computers in the Physical Chemistry Laboratory

The growth of computational chemistry and the ready availability of commercial ab initio packages has had a dramatic effect on the way that physical chemistry is practiced in the contemporary research laboratory. The clear implication is that without integration of computational chemistry into our physical chemistry laboratory curriculum we will be failing to teach our students how contemporary research is conducted. Fortunately, a number of approaches to including computational chemistry in the physical chemistry laboratory have been developed. These range from modifications of the full course to individual computational chemistry exercises for the laboratory. These developments can be found in Table VII.

Parish (*90*) developed a laboratory course in which four computational exercises were included along with the usual range of experiments. The exercises included symbolic programming and visualization using Maple, a molecular mechanics exercise in conformational searching, a comparison of molecular dynamics and Monte Carlo in exploring potential hypersurfaces, and finally a study of migration/elimination reactions using density functional theory (DFT) and ab initio methods. Another, even more integral approach has been taken by Schaertel, who coupled every experiment in her physical chemistry laboratory with an appropriate computational method (*91*). Gourley's "Theory and Experiment" course (*11*) also involves integral coupling of theoretical methods and experimentation.

In addition to these courses that integrate theory and experiment over several exercises, several individual exercises were developed that combined theory and experiments (*92-95*). The experimental methods ranged from cyclic voltammetry (*92*) to variable temperature NMR (*94*). The theoretical methods

Table VI. STM, AFM, Mass Spec, Cyclic Voltammetry and Miscellaneous Methods

Title	Ref. #
Electron Tunneling, a Quantum Probe for the Quantum World of Nanotechnology	72
Investigating Intermolecule Interactions via Scanning Tunneling Microscopy	73
Molecular Photography in the Undergraduate Laboratory: Identification of Functional Groups Using Scanning Tunneling Microscopy	74
Surface Oxidation Kinetics: A Scanning Tunneling Microscopy Experiment	75
Getting Physical with Your Chemistry: Mechanically Investigating Local Structure and Properties of Surfaces with the Atomic Force Microscope	76
Gaseous-Ion Fragmentation Mechanisms in Chlorobenzenes by GC/MS and GC/MS/MS: A Physical Chemical Approach for Undergraduates	77
Measuring Gas-Phase Basicities of Amino Acids Using an Ion Trap Mass Spectrometer	78
Design and Operation of a Portable Quadrupole Mass Spectrometer for the Undergraduate Curriculum	79
The Incorporation of a Single-Crystal X-ray Diffraction Experiment into the Undergraduate Physical Chemistry Laboratory	80
X-ray Diffraction Study of Alloys	81
Determination of the Real Surface Area of Pt Electrodes by Hydrogen Adsorption Using Cyclic Voltammetry	82
A Simple Student Experiment for Teaching Surface Electrochemistry: Adsorption of Polyoxometalate on Graphite Electrodes	83
Micelles in the Physical Chemistry Laboratory: Diffusion Coefficients and Half-Wave Potentials of Ferrocene	84
An Undergraduate Laboratory Experiment for the Direct Measurement of Monolayer-Formation Kinetics	85
A Low-Cost Matrix Isolation Experiment	86
Simulation and Analysis of Differential Scanning Calorimetry Output: Protein Unfolding Studies 1	87
Determination of a Diffusion Coefficient by Capillary Electrophoresis	88
ICP in the Physical Chemistry Laboratory: Determination of Plasma Temperature	89

included semiempirical force fields (*92, 93*), ab initio, DFT (*93, 94, 95*), and molecular mechanics methods (*95*).

Interestingly, in the experiments devoted solely to computational chemistry, molecular dynamics calculations had the highest representation (*96-98*). The method was used in simulations of simple liquids, (*96*), in simulations of chemical reactions (*97*), and in studies of molecular clusters (*98*). One experiment was devoted to the use of Monte Carlo methods to distinguish between first and second-order kinetic rate laws (*99*). One experiment used DFT theory to study two isomerization reactions (*100*).

Table VII. Computers in the Physical Chemistry Lab

Title	Ref. #
Introduction of Computational Chemistry into the Physical Chemistry Laboratory	*90*
Integrating Computational Chemistry into the Physical Chemistry Laboratory Curriculum: A Wet Lab/Dry Lab Approach	*91*
Using Cyclic Voltammetry and Molecular Modeling to Determine Substituent Effects in the One-Electron Reduction of Benzoquinones	*92*
Determination of the Rotameric Stability of 1,2-Dihaloethanes Using Infrared Spectroscopy	*93*
Rotational Barriers in Push-Pull Ethylenes: An Advanced Physical-Organic Project Including 2D EXSY and Computational Chemistry	*94*
The Energy Profile for Rotation about the C-C Bond in Substituted Ethanes	*95*
Molecular Dynamics Simulations of Simple Liquids	*96*
Molecular Dynamics Simulations of Chemical Reactions for Use in Education	*97*
Study of Molecular Clusters in Undergraduate Physical Chemistry	*98*
First-Order or Second-Order Kinetics? A Monte Carlo Answer	*99*
Computational Studies of Chemical Reactions: The HNC-HCN and CH_3NC-CH_3CN Isomerizations	*100*
Integration of National Instruments' LabVIEW Software into the Chemistry Curriculum	*101*
A Low-Cost Dynamic Surface Tension Meter with a LabVIEW Interface and Its Usefulness in Understanding Foam Formation	*102*
The Measurement of Activity Coefficients in Concentrated Electrolyte Solutions	*103*

In addition to computational chemistry, another use of computers in lab is the use of computers in the collection of experimental data. Two cited experiments (*101,102*) described the use of LabView in the physical chemistry laboratory, while the third involved computer data acquisition in an osmometry experiment (*103*).

Experiments on Modern or Relevant Topics

One of the points made in Schwenz and Moore was that the physical chemistry laboratory should better reflect the range of activities found in current physical chemistry research. This is reflected in part by the inclusion of modern instrumentation and computational methods, as noted extensively above, but also by the choice of topics. A number of experiments developed since Schwenz and Moore reflect these current topics. Some are devoted to modern materials, an extremely active research area, that I have broadly construed to include semiconductors, nanoparticles, self-assembled monolayers and other supramolecular systems, liquid crystals, and polymers. Others are devoted to physical chemistry of biological systems. I should point out here, that with rare exceptions, I have not included experiments for the biophysical chemistry laboratory in this latter category, primarily because the topics of many of these experiments fall out of the range of a typical physical chemistry laboratory or lecture syllabus. Systems of environmental interest were well represented as well.

Materials: Nanoparticles, Self-Assembled Monolayers & Supramolecular Systems, Liquid Crystals and Polymers

A number of experiments were devoted to semiconductors (*104*), and semiconductor nanoparticles (*38, 105-109*). Some of the experiments were devoted to the synthesis or growth process of the nanoparticles (*105, 106, 109*), while others were devoted to the unique properties of the particles (*107, 108, 109*).

Another well-represented category was that of self-assembled monolayers (SAMS) and other supramolecular systems. The experiments on the SAMS included studies of the surface pKa of one system (*110*), the kinetics and thermodynamics of the self-assembly process (*111*), and the characterization of the SAM surface by study of solution contact angles (*112*). The experiments on supramolecular systems included studies on chemical equilibria in such systems (*113, 114, 115*), the kinetics of inclusion phenomena (*116*), and the use of solvatochromic probes in studying inclusion phenomena (*117*).

One experiment on liquid crystals was a multiweek effort devoted to synthesis of the material, and study of several of the important characteristics of the liquid crystals (*118*). This was in some ways a model paper, because in addition to describing an interesting project, it also provided a clear pedagogical rationale for the experiment, placed it in context in the curriculum, and included an assessment of how well the goals for the project were met. Other studies of liquid crystals included a computational approach to their study (*119*), and the study of rotational mobility in these unique materials (*120*).

Several experiments were devoted to the study of polymers. One was devoted to the study of spin-coating (*121*). A second uses light scattering to measure polymer sizes (*122*). There were several approaches to polymer kinetics (*39, 123, 124, 125*). Finally there was one study using differential scanning calorimetry to study polymer glasses (*126*). All of these experiments on modern materials can be found listed in Table VIII.

Current Topics: Physical Chemistry of Biological Systems

A small number of experiments that are listed in Table IX were devoted to the physical chemistry of biological systems. The small number is surprising given the number of elegant experiments in the chemical literature in which physical chemical methods have been applied to biological problems. Nonetheless the range of experiments was very interesting. They included measurement of quantum yields and other parameters in a photosynthetic O_2 evolving system (*127*), the study of fluorescence in two proteins (*128*), measurement of renaturation rates of DNA (*129*), the use of viscosity measurements to infer the shapes of macromolecules (*130*), and kinetic studies on the removal of metal centers from enzymes (*131*). In addition, three experiments studied systems that could be used as models for biological systems. Two of these were studies of the effects of micelles on equilibria (*132*) or reaction rates (*133*). The third studied hydrogen bonding equilibria in a multi-phase system, an attempt to model the role of hydrogen bonds in biological systems where both hydrophobic and hydrophilic regions are present (*134*).

Current Topics: Environmental and Miscellaneous Current Topics

Environmental issues are a current topic that strongly motivates many students and that gets much attention in the news. Several new experiments were devoted to environmental topics. We have already mentioned a laser lab devoted to the study of the kinetics of ozone formation (*26*). Other new experiments studied the thermochemistry of biodiesel (*135*), the use of spectroscopy to estimate the average temperature of the earth's atmosphere

Table VIII. Experiments on Modern Materials

Title	Ref. #
The Dependence of Resistance on Temperature for Metals, Semiconductors and Superconductors	104
A Safer, Easier, Faster Synthesis for CdSe Quantum Dot Nanocrystals	105
Growth Kinetics and Modeling of ZnO Nanoparticles	106
Quantum Dots: An Experiment for Physical or Materials Chemistry	107
Photoelectrochromism in Tungsten Trioxide Colloidal Solutions	108
Synthesis and Optical Properties of Quantum-Size Metal Sulfide Particles in Aqueous Solution	109
Surface pK_a of Self-Assembled Monolayers	110
Kinetics and Thermodynamics of Au Colloid Monolayer Self-Assembly	111
Contact Angles of Aqueous Solutions on Copper Surfaces Bearing Self-Assembled Monolayers	112
Chemical Equilibrium in Supramolecular Systems as Studied by NMR Spectroscopy	113
A Practical Integrated Approach to Supramolecular Chemistry. I. Equilibria in Inclusion Phenomena.	114
Determination of Thermodynamic Processes of the Cyclodextrin Inclusion Process	115
A Practical Integrated Approach to Supramolecular Chemistry. II. Kinetics of Inclusion Phenomena	116
Cyclodextrin Inclusion Complexes with a Solvatochromic Fluorescent Probe	117
Synthesis and Physical Properties of Liquid Crystals: An Interdisciplinary Experiment	118
A Computer Experiment in Physical Chemistry: Linear Dichroism in Nematic Liquid Crystals	119
Rotational Mobility in a Crystal Studied by Dielectric Relaxation Spectroscopy	120
Spin-Coating of Polystyrene Thin Films as an Advanced Undergraduate Experiment	121
Spectrofluorimeters as Light-Scattering Apparatus: Application to Polymers Molecular Weight Determination	122
Use of a Reliable Homemade Dilatometer To Study the Kinetics of the Radical Chain Polymerization of PMMA	123
Polystyrene Kinetics by Infrared	124
Propagating Fronts of Polymerization in the Physical Chemistry Laboratory	125
Study of Polymer Glasses by Modulated Differential Scanning Calorimetry in the Undergraduate Physical Chemistry Laboratory	126

(*136*), studies of carbon dioxide concentration relaxation (*137*), the adsorption of p-nitrophenol from an aqueous solution (a model system for remediation of polluted water) (*138*), and finally a study of the use of semiconductors in solar energy conversion (*139*).

There were also several experiments involving current topics that did not fit neatly into a category. These included studies of supercritical phenomena (*140, 141*), cometary spectroscopy (*142*), experimental studies on the thermodynamics of heat pumps (*143*), a study of industrially relevant phase transfer catalysts (*144*), and finally an electrochemical study of a commercial soap (*145*). Both the environmental and the miscellaneous experiments can be found listed in Table X.

Integrated Laboratories

Another modern trend is the development of laboratory courses based on multi-week projects, involving not just one subdiscipline of chemistry but two or more. Such laboratory courses have been labeled "Integrated Laboratories". They are motivated in part by the belief that this type of integrated approach more closely models the way real chemical research is done and will provide stronger motivation for students than the more traditional one-subject laboratory courses.

One example of such a course is the Haverford "Laboratory in Chemical Structure and Reactivity" (*146*) that includes six projects, each of which involves sample preparation, sample analysis and some kind of determination of the properties of the substance prepared. The projects include organopalladium chemistry, porphyrin photochemistry, enantioselective synthesis, computer-aided modeling, enzyme kinetics and electron transfer reactions.

Another example is the interdisciplinary laboratory developed at Harvey Mudd College (*147*) in which eight different interdisciplinary experiments, ranging from thermal properties of an ectothermic animal to synthesis and characterization of liquid crystals, are carried out over two semesters.

In addition to development of full laboratory courses, several individual experiments that can be included in such integrated laboratories have been developed recently. They include experiments on photocatalysis (*148*), synthesis, kinetics and thermodynamics of an inorganic compound (*149*), studies on conformational analysis (*150*), synthesis and variable temperature proton NMR of an inorganic compound (*151*), and the study of microemulsions (*152*). As such laboratories become more common, we can expect more of these experiments to appear in the literature. These integrated laboratory courses and experiments can be found in Table XI.

Table IX. Physical Chemistry of Biological Systems

Title	Ref. #
Measurement of Quantum Yield, Quantum Requirement, and Energetic Efficiency of the O_2-Evolving System of Photosynthesis by a Simple Dye Reaction	127
Ribonuclease T_1 and Alcohol Dehydrogenase Fluorescence Quenching by Acrylamide	128
Examining the Rate of Renaturation for Genomic DNA Isolated from E. Coli.	129
Viscosity and the Shapes of Macromolecules	130
Removal of Zinc from Carbonic Anhydrase	131
Micelles in the Physical/Analytical Chemistry Laboratory. Acid Dissociation of Neutral Red Indicator	132
Micellar Effects on the Spontaneous Hydrolysis of Phenyl Chloroformate	133
A Family of Hydrogen Bonds in the Model System Salicylic Acid – Toluene – Water	134

Table X. Environmental Experiments and Miscellaneous Topics

Title	Ref. #
Determination of the Heat of Combustion of Biodiesel Using Bomb Calorimetry	135
Fine-Structure Measurements of Oxygen A Band Absorbance for Estimating the Thermodynamic Average Temperature of the Earth's Atmosphere	136
Carbon Dioxide Dissolution as a Relaxation Process	137
Adsorption of p-Nitrophenol from Dilute Aqueous Solution	138
Photoelectroconversion by Semiconductors	139
Exploring Phase Diagrams Using Supercritical Fluids	140
Measuring P-V-T Phase Behavior with a Variable Volume View Cell	141
Cometary Spectroscopy for Advanced Undergraduates	142
Experimental Demonstration of the Principles of Thermal Energy Storage and Chemical Heat Pumps	143
Phase-Transfer Catalytic Reactions	144
An Undergraduate Physical Chemistry Experiment on Surfactants: Electrochemical Study of a Commercial Soap	145

Table XI. Integrated Laboratories and Experiments for Integrated Laboratories

Title	Ref. #
Laboratory in Chemical Structure and Reactivity – Superlab II	146
An Integration of Chemistry, Biology and Physics: The Interdisciplinary Laboratory	147
Photocatalysis, A Laboratory Experiment for an Integrated Physical Chemistry-Instrumental Analysis Course	148
Synthesis, Kinetics, and Thermodynamics: An Advanced Laboratory Investigation of the Cis-Trans Isomerization of $Mo(CO)_4(PR_3)_2$	149
Conformational Analysis in an Advanced Integrated Laboratory Course	150
Synthesis and Variable-Temperature ^1H NMR Conformational Analysis of Bis(η^5-cyclopentadienyl)titanium Pentasulfide	151
Microemulsions as a New Working Medium in Physical Chemistry	152

Traditional Experiments

In addition to the activity documented above there has been a tremendous amount of activity in the development of more traditional experiments for the physical chemistry laboratory. Some of these experiments are improvements on older methods, some involve new systems, and some involve new types of analysis. There are far too many of these experiments to discuss individually, but all of them will be found listed in tables below. They have been divided roughly into spectroscopy and the electronic structure of matter, thermodynamics, including thermochemistry and properties of liquids, solids and solutions, and kinetics, including photochemistry.

The spectroscopy experiments are further subdivided into: atomic spectroscopy found in Table XII, infrared and Raman spectra found in table XIII, visible and ultraviolet absorption spectroscopy found in table XIV, and luminescence spectroscopies found in table XV.

The thermodynamics experiments are subdivided into experiments on calorimetry and heat capacity, Table XVI; phase transitions, Table XVII; properties of gases, liquids, solids, solutions and mixtures, Table XVIII; and finally equilibrium and miscellaneous "thermodynamic topics", Table XIX.

The kinetics experiments are subdivided into classical kinetics experiments, Table XX; photochemistry, Table XXI; catalysis, oscillating reactions and miscellaneous topics in kinetics, Table XXII.

Table XII. Atomic Spectroscopy Experiments

Title	Ref. #
Use of the Vreeland Spectroscope in the Quantum Chemistry Laboratory	153
The Balmer Spectrum of Hydrogen: An Old Experiment with a New Twist	154
A Procedure to Obtain the Effective Nuclear Charge from the Atomic Spectrum of Sodium	155
The Ionization Energy of Helium	156
Mass Ratio of the Deuteron and Proton from the Balmer Spectrum of Hydrogen	157

Table XIII. Infrared and Raman Spectroscopy Experiments

Title	Ref. #
Geometry of Benzene from the Infrared Spectrum	158
Determination of Surface Coverage of an Adsorbate on Silica Using FTIR Spectroscopy	159
Infrared Spectrum and Group Theoretical Analysis of the Vibrational Modes of Carbonyl Sulfide	160
Applications of Group Theory: Infrared and Raman Spectra of the Isomers of 1,2-Dichloroethylene	161
Modeling Stretching Modes of Common Organic Molecules with the Quantum Mechanical Harmonic Oscillator	162
Isotope Ratio, Oscillator Strength, and Band Positions from CO_2 IR Spectra	163
The Fundamental Rotational-Vibrational Band of CO and NO	164
Vibrational-Rotational Spectra: Simultaneous Generation of HCl, DCl, HBr and DBr	165
Determination of the Rotational Barrier in Ethane by Vibrational Spectroscopy and Statistical Thermodynamics	166
Using the Asymmetric Stretch Band of Atmospheric CO_2 to Obtain the C=O Bond Length	167
An Interactive Dry Lab Introduction to Vibrational Raman Spectroscopy Using Carbon Tetrachloride	168

Table XIV. Visible/Ultraviolet Absorption Spectroscopy Experiments

Title	Ref. #
Cotton Effect in Copper-Proline Complexes in the Visible Region	169
The Electronic Absorption Spectrum of Molecular Iodine: A New Fitting Procedure for the Physical Chemistry Laboratory	170
Influence of Dielectric Constant on the Spectral Behavior of Pinacyanol	171
Use of Hückel Molecular Orbital Theory in Interpreting the Visible Spectra of Polymethine Dyes	172
An Experiment in Electronic Spectroscopy: Information Enhancement Using Second Derivative Analysis	173
Visible Spectra of Conjugated Dyes: Integrating Quantum Chemical Concepts with Experimental Data	174
Alternative Compounds for the Particle-in-a-Box Experiment	175
Visible Absorption Spectroscopy and Structure of Cyanine Dimers in Aqueous Solution	176

Table XV. Luminescence Spectroscopy Experiments and Miscellaneous Electronic Structure Experiments

Title	Ref. #
A Simple Device to Demonstrate the Principles of Fluorometry	177
Micellar Aggregation Numbers – A Fluorescence Study	178
Fluorescence and Light Scattering	179
Are Fluorescence Quantum Yields So Tricky to Measure? A Demonstration Using Familiar Stationery Products	180
Fluorescence Measurement of Pyrene Wall Adsorption and Pyrene Association with Humic Acids	181
Spectroscopy of Flames: Luminescence Spectra of Reactive Intermediates	182
Fluorescence Quenching in Microheterogeneous Media	183
Enhanced Luminescence of Lanthanides: Determination of Europium by Enhanced Luminescence	184
Luminescence Quenching in Microemulsion Studies	185
A Room-Temperature Emission Lifetime Experiment for the Physical Chemistry Laboratory	186
Phosphorescent Lifetimes and Spectra	187
Molecular Photophysics of Acridine Yellow Studied by Phosphorescence and Delayed Fluorescence	188
An Experimental Determination of the Second Radiation Constant	189
Cost-Effective Spectroscopic Instrumentation for the Physical Chemistry Laboratory	190
Determination of the Magnetic Moments of Transition Metal Complexes Using Rare Earth Magnets	191

Table XVI. Calorimetry and Heat Capacity Experiments

Title	Ref. #
Calorimetric Determination of Aqueous Ion Enthalpies	192
An Integrated-Circuit Temperature Sensor for Calorimetry and Differential Temperature Measurement	193
Simultaneous Determinations of Sulfur and Heat Content of Coal	194
The Isothermal Heat Conduction Calorimeter: A Versatile Instrument for Studying Processes in Physics, Chemistry, and Biology	195
LabWorks and the Kundt's Tube: A New Way To Determine the Heat Capacities of Gases	196
Thermal Analysis of Carbon Allotropes	197
Thermoprogrammed Desorption	198

Table XVII. Experiments on Phase Transitions

Title	Ref. #
Determination of Enthalpy of Vaporization Using a Microwave Oven	199
Enthalpy of Vaporization by Gas Chromatography	200
Vapor Pressure Measurements in a Closed System	201
A Novel Method for Examination of Vapor-Liquid Equilibria	202
Determination of Enthalpy of Vaporization of Pure Liquids by UV Spectrometry	203
A Convenient, Inexpensive, and Environmentally Friendly Method of Measuring the Vapor Pressure of a Liquid as a Function of Temperature	204
The Solid-Liquid Phase Diagram Experiment	205
Multichannel DTA Apparatus for the Measurement of Phase Diagrams	206

Table XVIII. Experiments on Properties of Gases, Liquids; Solids, Solutions and Mixtures

Title	Ref. #
Diffusion of Water Vapor	207
Rare Gas Viscosities: A Learning Tool	208
Application of Light-Emitting Diodes and Photodiodes Coupled to Optical Fibers To Study the Dependence of Liquid Viscosity on Temperature	209
Quantifying Meniscus Forces with an Electronic Balance: Direct Measurement of Liquid Surface Tension	210
An-Easy-To-Build Rotational Viscometer with Digital Readout	211
A Multipurpose Apparatus to Measure Viscosity and Surface Tension of Solutions	212
Mechanical Coefficients of Liquids by a Differential Expansion Method	213
Convection in a Continuously Stratified Fluid	214
Raoult's Law: Binary Liquid-Vapor Phase Diagrams	215
The Vapor Pressure of Liquid Binary Solutions	216
Survey of Binary Azeotropes as Physical Chemistry Laboratory Experiments with Attention to Cost, Safety, and the Environment	217
A Physical Chemistry Lab Project: The Effect of Composition on Several Physical Properties of Binary Mixtures of Common Liquids	218
A Laboratory Method for the Determination of Henry's Law Constants of Volatile Organic Chemicals	219
An Introductory Experience for Physical Chemistry: Victor Meyer Revisited	220
Contact Angle Goniometry as a Tool for Surface Tension Measurements of Solids, Using Zisman Plot Method	221

Table XIX. Experiments on Equilibrium and Miscellaneous Thermodynamic Topics

Title	Ref. #
Iron(III) Thiocyanate Revisited: A Physical Chemistry Equilibrium Lab Incorporating Ionic Strength Effects	222
A Dibasic Acid Titration for the Physical Chemistry Laboratory	223
The Effect of Diffusion Due to a Temperature Gradient on the Measurement of the N_2O_4-NO_2 Equilibrium Constant	224
Simultaneous Determination of the Ionization Constant and the Solubility of Sparingly Soluble Drug Substances	225
Determination of the Critical Micelle Concentration of Cationic Surfactants	226
The Solubility Product of $PbCl_2$ from Electrochemical Measurements	227
Rubber Elasticity: A Simple Method for Measurement of Thermodynamic Properties	228

Table XX. Classical Kinetics Experiments

Title	Ref. #
Gas Phase Kinetics of $O_3 + C_2H_2$ and $O_3 + H_2C=CCl_2$	229
Graphical Interface for the Study of Gas-Phase-Reaction Kinetics: Cyclopentane Vapor Pyrolysis	230
The Spontaneous Hydrolysis of Methyl Chloroformate	231
Hydrolysis of Pentaamminechlorocobalt(III): A Unified Mechanistic View	232
Kinetics of Methylene Blue Reduction by Ascorbic Acid	233
Decomposition of Aspartame	234
Synthesis and Kinetics of Hydrolysis of 3,5-Dimethyl-N-acetyl-p-benzoquinone Imine	235
Chemical Kinetics in Real Time: Using the Differential Rate Law and Discovering the Reaction Orders	236
Pressure Measurements to Determine the Rate Law of the Magnesium-Hydrochloric Acid Reaction	237
Evaporation Kinetics in Short-Chain Alcohols by Optical Interference	238
A Novel Experiment in Chemical Kinetics: The A $\Leftrightarrow$B$\rightarrow$C Reaction System	239
An Undergraduate Physical Chemistry Experiment on the Analysis of First-Order Kinetic Data	240
Ionic Strength Effect on the Rate of Reduction of Hexacyanoferrate(III) by Ascorbic Acid	241
A Fluorimetric Approach to Studying the Effects of Ionic Strength on Reaction Rates	242
Effect of Dielectric Constant and Ionic Strength on the Fading of N,N-Dimethylaminophenolphthalein in Alkaline Medium	243
Kinetics of Reduction of Toluidine Blue with Sulfite – Kinetic Salt Effect in Elucidation of Mechanism	244
Micellar Effects on the Spontaneous Hydrolysis of Phenyl Chloroformate	245
Micelles in the Physical Chemistry Laboratory. Kinetics of Hydrolysis of 5,5'-Dithiobis-(2-nitrobenzoate)	246
Kinetic Solvent Isotope Effect: A Simple Multipurpose Physical Chemistry Experiment	247

Table XXI. Photochemistry Experiments

Title	Ref. #
Photochemistry of Benzophenone in 2-Propanol	248
Photochemistry of Chloropicrin	249
Heterogeneous Photochemistry	250
Chemical Actinometry: Using o-Nitrobenzaldehyde to Measure Light Intensity in Photochemical Experiments	251
An Experiment on Photochromism and Kinetics for the Undergraduate Laboratory	252
An Easy and Inexpensive Flash Spectroscopy Experiment	253

Table XXII. Experiments on Catalysis, Oscillating Reactions and Miscellaneous Topics in Kinetics

Title	Ref. #
Catalytic Oxidation of Sulfur Dioxide by Activated Carbon	254
A Kinetics Experiment to Demonstrate the Role of a Catalyst in a Chemical Reaction	255
Spectroscopic Monitoring of the Heterogeneous Catalytic Decomposition of Gaseous Ammonia	256
Chemical Oscillations and Waves in the Physical Chemistry Lab	257
Nonlinear Dynamics of the BZ Reaction: A Simple Experiment that Illustrates Limit Cycles, Chaos, Bifurcations and Noise	258
The BZ Reaction: Experimental and Model Studies in the Physical Chemistry Laboratory	259
A Stopped-Flow Kinetics Experiment for Advanced Undergraduate Laboratories: Formation of Iron(III) Thiocyanate	260
A Simple Electrochemical Approach to Heterogeneous Reaction Kinetics	261
A Student Experiment in Non-Isothermal Chemical Kinetics	262

Conclusions

The development of physical chemistry experiments is an active and vibrant endeavor. Since Schwenz and Moore came out in 1993, over 250 new experiments have been developed. These experiments involved a wide range of modern instrumentation and modern topics. They included innovative pedagogical approaches. They included a wide range of traditional topics as well. Together they reflect the importance of a wide range of experiments and approaches in a healthy physical chemistry curriculum.

However, there were some notable lacunae that need to be addressed. Little work has been done in chemical education research on the physical chemistry laboratory, although what has been done is both valuable and excellent. In addition, little attention has been paid to the issue of the structure of the physical chemistry laboratory as a whole (or at least little has been published). More needs to be done in this area. Very few of the experiments included a clear pedagogical objective, and those that did, failed to do any assessment of those objectives. It is hoped that a continuously increasing percentage of new experiments will include these elements.

In the development of modern experiments, there were a large number of experiments developed using lasers and the NMR. More development would be welcome in experiments using MS, AFM, and STM. In addition, more experiments devoted to characteristics of modern materials, to environmental chemistry, and especially to the physical chemistry of biological and biologically relevant systems are needed.

In short, the state of the physical chemistry laboratory mirrors that of chemistry itself. It is vibrant and exciting, with many solid achievements in hand, but with wide frontiers for development and exploration. The next 13 years are ones to look forward to.

References

1. *Physical Chemistry: Developing A Dynamic Curriculum*; Schwenz, Richard W; Moore, Robert, J., Eds.; American Chemical Society: Washington, DC, 1993
2. MacCay, Colin F. In *Physical Chemistry: Developing a Dynamic Curriculum*; Schwenz, Richard W; Moore, Robert, J., Eds.; American Chemical Society: Washington, DC, 1993, pp. 74-83.
3. *How People Learn: Brain, Mind, Experience and School*; Bransford, John D.; Brown, Ann L.; Cocking, Rodney R.; Donovan, M. Suzanne; Pellegrino, James W., Eds.; National Academy Press, Washington, DC, 2000
4. *How Students Learn: History, Mathematics, and Science in the Classroom*; Donovan, M. Suzanne; Bransford, John D., Eds; National Academies Press, Washington, DC, 2005
5. Malina, Eric G.; Nakleh, Mary B. *J. Chem. Educ.* **2003**, *80*, 691-698.
6. Unpublished, Weaver, Gabriela C.; Department of Chemistry, Purdue University
7. Long, George; Howald, Reed; Miderski, Carol Ann; Zielinkski, Teresa Julia. *Chem. Educator* **1996**, *1*, 1-17

8. Slocum, Laura E.; Towns, Marcy Hamby; Zielinski, Theresa Julia. *J. Chem. Educ.* **2004**, *81*, 1058-1065.

9. Sauder, Deborah; Towns, Marcy; Derrick, Betty; Grushow, Alexander; Kahlow, Michael; Long, George; Miles, Danny; Shalhoub, George; Stout, Roland; Vaksman, Michael; Pfeiffer, William F.; Weaver, Gabriella; Zielinski, Theresa Julia. *Chem. Educator* **2000**, *5*, 77-82.

10. Deckert, Alice A.; Nestor, Lisa P.; DiLullo, Donna. *J. Chem. Educ.* **1998**, *75*, 860-863.

11. Unpublished, Gourley, Bridget, L.; Department of Chemistry, DePauw University

12. Long, George; Sauder, Deborah; Shalhoub, George M.; Stout, Roland; Towns, Marcy Hamby; Zielinkski, Theresa Julia. *J. Chem. Educ.* **1999**, *76*, 841-847.

13. Buckley, Paul D.; Jolley, Kenneth W; Watson, Ian D. *J. Chem. Educ.* **1997**, *74*, 549-551

14. Muenter, John S. *J. Chem. Educ.* **1996**, *73*, 576-580.

15. Tran, Yang; Whitten, J. E. *J. Chem. Educ.* **2001**, *78*, 1093-1095.

16. Van Dyke, David A.; Pryor, Brian A.; Smith, Philip G.; Topp, Michael R. *J. Chem. Educ.* **1998**, *75*, 615-620.

17. Muenter, John S.; Deutsch, John L. *J. Chem Educ.* **1996**, *73*, 580-585.

18. Gsponer, Héctor E.; Argüello, Gustavo A.; Argüello, Gerardo A. J. Chem. Educ. 1997, 74, 968-972.

19. Lasher, D.P.; DeGraff, B. A.; Augustine, B. H. *J. Chem. Educ.* **2000**, *77*, 1201-1203.

20. Gutow, Jonathan H. *J. Chem. Educ.* **2005**, *82*, 302-305.

21. Henderson, Giles; Tennis, Ronald; Ramsey, Terry. *J. Chem. Educ.* **1998**, *75*, 1139-1142.

22. Salter, Carl; Range, Kevin; Salter, Gail. *J. Chem. Educ.* **1999**, *76*, 84-85.

23. Windisch, Charles F., Jr.; Exarhos, Gregory J.; Sharma, Shiv K. *J. Chem. Educ.* **2005**, *82*, 916-918.

24. Poulsen, Lars; Ruiz, Arantza Zabala; Pedersen, Steen Uttrup; Ogilby, Peter R. *J. Chem. Educ.* **2003**, *80*, 819-821.

25. Masiello, Tony; Vulpanovici, Nicolae; Nibler, Joseph W. *J. Chem. Educ.* **2003**, *80*, 914-916.

26. Krasnoperov, Lev N.; Stepanov, Victor. *J. Chem. Educ.* **1999**, *76*, 1182-1183.

27. Bengali, Ashfaq A.; Charlton, Samantha B. *J. Chem. Educ.* **2000**, *77*, 1348.

28. DeGraff, Benjamin A.; Horner, David A. *J. Chem. Educ.* **1996**, *73*, 279-285.

142

29. Fitzwater, David A.; Thomasson, Kathryn A.; Glinski, Robert J. *J. Chem. Educ.* **1995**, *72*, 187-189.
30. Young, Mark D.; Borjemscaia, Natalia C.; Wladkowski, Brian D. *J. Chem. Educ.* **2005**, *82*, 912-915.
31. Comstock, Matthew G; Gray, Jeffrey A. *J. Chem. Educ.* **1999**, *76*, 1272-1275.
32. Nissum, Mikkel; Shabanova, Elizabeth; Nielsen, Ole Faurskov. *J. Chem. Educ.* **2000**, *77*, 633-637.
33. Melin, Stéphanie; Nibler, Joseph W. *J. Chem. Educ.* **2003**, *80*, 1187-1190
34. Grant, Christopher A; Hardwick, J. L. *J. Chem. Educ.* **1997**, *74*, 318-321.
35. Weaver, Gabriela C.; Norrod, Karen. *J. Chem. Educ.* **1998**, *75*, 621-624.
36. Stenland, Chris; Pettitt, B. Montgomery. *J. Chem. Educ.* **1995**, *72*, 560-564.
37. Miles, Danny G., Jr.; Yang, Zhihao; Yu Hyuk. *J. Chem. Educ.* **2002**, *79*, 1007-1012.
38. Ahn, Heejoon; Whitten, James E. *J. Chem. Educ.* **2005**, *82*, 909-911.
39. Page, Melissa A.; Grubbs, W. Tandy. *J. Chem. Educ.* **1999**, *76*, 666-668.
40. Sattar, Simeen; Rinehart, Frank P. *J. Chem. Educ.* **1998**, *75*, 1136-1138.
41. Munguia, Teresita; Smith, Charles A. *J. Chem. Educ.* **2001**, *78*, 343-344.
42. Urian, R. Craig; Khundkar, Lutfur R. *J. Chem. Educ.* **1998**, *75*, 1135.
43. Fletcher, Beth; Grabowski, Joseph J. *J. Chem. Educ.* **2000**, *77*, 640-645.
44. Shaw, Roosevelt; Hokmabadi, Mohammad. *J. Chem. Educ.* **1996**, *73*, 474-475.
45. Riley, Scott A.; Franklin, Nathan R.; Oudinarath, Bobbie; Wong, Sally; Congalton, David; Nishimura, A. M. *J. Chem. Educ.* **1997**, *74*, 1320-1322.
46. Seidman, Kurt; Payne, Amy. *J. Chem. Educ.* **1998**, *75*, 897-900.
47. O'Brien, L.C.; Kubicek, R. L. *J. Chem. Educ.* **1996**, *73*, 86-87.
48. Bradley, Michael S.; Bratu, Cheryl. *J. Chem. Educ.* **1997**, *74*, 553-555.
49. Comstock, Matthew G.; Kerr, Jeffrey R.; Gray, Jeffrey A. *J. Chem. Educ.* **2002**, *79*, 500-502.
50. Vaksman, Michael A.; Lane, James W. *J. Chem. Educ.* **2001**, *78*, 1507-1509.
51. Whitten, J. E. *J. Chem. Educ.* **2001**, *78*, 1096-1100.
52. Williams, Kathryn R.; Adhyaru, Bhavin; German, Igor; Alvarez, Eric. *J. Chem. Educ.* **2002**, *79*, 372-373.
53. Emerson, David W.; Steinberg, Spencer M.; Titus, Richard L. *J. Chem. Educ.* **2005**, *82*, 466-467.

54. Grushow, Alexander; Zielinski, Theresa Julia. *J. Chem. Educ.* **2002**, *79*, 707-714.

55. Lessinger, Leslie. *J. Chem. Educ.* **1995**, *72*, 85-87

56. Peters, Steven J.; Stevenson, Cheryl D. *J. Chem. Educ.* **2004**, *81*, 715-717.

57. Potts, Richard A.; Schaller, Ruth A. *J. Chem. Educ.* **1993**, *70*, 421-424.

58. Nibler, Joseph W.; Minarik, Philip; Fitts, William; Kohnert, Rodger. *J. Chem. Educ.* **1996**, *73*, 99-101.

59. Morris, Kevin F.; Erickson, Luther E. *J. Chem. Educ.* **1996**, *73*, 471-473.

60. Jarek, Russell L.; Flesher, Robert J.; Shin, Seung Koo. *J. Chem. Educ.* **1997**, *74*, 978-982.

61. Ba, Yang; Han, Steven; Ni, Lily; Su, Tony; Garcia, Andres. *J. Chem. Educ.* **2006**, *83*, 296-298.

62. Brown, Keith C.; Tyson, Randy L.; Weil, John A. *J. Chem. Educ.* **1998**, *75*, 1632-1635.

63. Jameson, Donald L.; Anand, Rajan. *J. Chem. Educ.* **2000**, *77*, 88-89.

64. Lorigan, Gary A.; Minto, Robert E.; Zhang, Wei. *J. Chem. Educ.* **2001**, *78*, 956-958.

65. Gasyna, Zbigniew L.; Jurkiewicz, Antoni. *J. Chem. Educ.* **2004**, *81*, 1038-1039.

66. Nestle, Nikolaus; Dakkouri, Marwan; Rauscher, Hubert. *J. Chem. Educ.* **2004**, *81*, 1040.

67. Gaede, Holly C.; Stark, Ruth E. *J. Chem. Educ.* **2001**, *78*, 1248-1250.

68. Zanger, Murray; Moyna, Guillermo. *J. Chem. Educ.* **2005**, *82*, 1390-1392.

69. Mills, Nancy S. *J. Chem. Educ.* **1996**, *73*, 1190-1192.

70. Basu, Partha *J. Chem. Educ.* **2001**, *78*, 666-669.

71. Butera, R. A.; Waldeck, D. H. *J. Chem. Educ.* **2000**, *77*, 1489-1491.

72. Hipps, K. W.; Scuderio, L. *J. Chem. Educ.* **2005**, *82*, 704-711.

73. Pullman, David; Peterson, Karen I. *J. Chem. Educ.* **2004**, *81*, 549-552.

74. Giancarlo, Leanna C.; Fang, Hongbin; Avila, Luis; Fine, Leonard W.; Flynn, George W. *J. Chem. Educ.* **2000**, *77*, 66-71.

75. Poler, Jordan C. *J. Chem. Educ.* **2000**, *77*, 1198-1200.

76. Heinz, William F.; Hoh, Jan H. *J. Chem. Educ.* **2005**, *82*, 695-703.

77. Schildcrout, Steven M. *J. Chem. Educ.* **2000**, *77*, 501-502.

78. Sunderlin, Lee S.; Ryzhov, Victor; Keller, Lanea M. M.; Gaillard, Elizabeth R. *J. Chem. Educ.* **2005**, *82*, 1071-1073.

79. Henchman, Michael; Steel, Colin. *J. Chem. Educ.* **1998**, *75*, 1042-1049.

80. Crundwell, G.; Phan, J.; Kantarjieff, K. A. *J. Chem. Educ.* **1999**, *76*, 1242-1245.

81. Butera, R. A.; Waldeck, D. W. *J. Chem. Educ.* **1997**, *74*, 115-119.

82. Rodríguez, José M. Doña; Melián, José Alberto Herrera; Peña, Jesús Pérez. *J. Chem. Educ.* **2000**, *77*, 1195-1197.

83. Martel, David; Sojic, Neso; Kuhn, Alexander. *J. Chem. Educ.* **2002**, *79*, 349-352.

84. Williams, Kathryn R.; Bravo, Roberto. *J. Chem. Educ.* **2000**, *77*, 392-394.

85. Karpovich, D. S.; Blanchard, G. J. *J. Chem. Educ.* **1995**, *72*, 466-470.

86. Flair, Mark; Fletcher, T. Rick. *J. Chem. Educ.* **1995**, *72*, 753-755.

87. Chowdhry, Babur; Leharne, Stephen. *J. Chem. Educ.* **1997**, *74*, 236-240.

88. Williams, Kathryn R.; Adhyaru, Bhavin; German, Igor; Russell, Thomas. *J. Chem. Educ.* **2002**, *79*, 1475-1476.

89. Page, Melissa A.; Smith, Benjamin W.; Williams, Kathryn R. *J. Chem. Educ.* **2002**, *79*, 364.

90. Unpublished. Parish, Carol A. Hobart and William Smith Colleges. Current address: University of Richmond, Richmond, VA.

91. Karpen, Mary E.; Henderleiter, Julie; Schaertel, Stephanie A. *J. Chem. Educ.* **2004**, *81*, 475-477.

92. Heffner, Janell E.; Raber, Jeffrey C.; Moe, Owen A.; Wigal, Carl T. *J. Chem. Educ.* **1998**, *75*, 365-367.

93. Wladkowski, Brian D.; Broadwater, Steven J. *J. Chem. Educ.* **2002**, *79*, 230-233.

94. Dwyer, Tammy J.; Normal, Julia E.; Jasien, Paul G. *J. Chem. Educ.* **1998**, *75*, 1635-1640.

95. Erickson, Luther E.; Morris, Kevin F. *J. Chem. Educ.* **1998**, *75*, 900-906.

96. Speer, Owen F.; Wengerter, Brian C.; Taylor, Ramona S. *J. Chem. Educ.* **2004**, *81*, 1330-1332.

97. Xie, Qian; Tinker, Robert. *J. Chem. Educ.* **2006**, *83*, 77-83.

98. Kahn, D.; Viswanathan, R. *J. Chem. Educ.* **1997**, *74*, 982-984.

99. Tellinghuisen, Joel. *J. Chem. Educ.* **2005**, *82*, 1709-1714.

100. Halpern, Arthur A. *J. Chem. Educ.* **2006**, *83*, 69-76.

101. Drew, Steven M. *J. Chem. Educ.* **1996**, *73*, 1107.

102. Spanoghe, Pieter; Cocquyt, Jan; Van der Meeren, Paul. *J. Chem. Educ.* **2001**, *78*, 338-342.

103. Bonicamp, Judith M.; Loflin, Ashley; Clark, Roy W. *J. Chem. Educ.* **2001**, *78*, 1541-1543.

104. Butera, R. A.; Waldeck, D. H. *J. Chem. Educ.* **1997**, *74*, 1090-1094.

105. Boatman, Elizabeth M.; Lisensky, George C.; Nordell, Karen J. J. Chem. Educ. 2005, 82, 1697-1699.

106. Hale, Penny S.; Maddox, Leone M.; Shapter, Joe G.; Voelcker, Nico H.; Ford, Michael J.; Waclawik, Eric R. *J. Chem. Educ.* **2005**, *82*, 775-778.

107. Winkler, L. D.; Arceo, J. F.; Hughes, W. C.; DeGraff, B. A.; Augustine, B. H. *J. Chem. Educ.* **2005**, *82*, 1700-1702.

108. Chenthamarakshan, C. R.; de Tacconi, N. R.; Xu, Lucy; Rajeshwar, Krishnan. *J. Chem. Educ.* **2004**, *81*, 1790-1793.

109. Nedeljković, J. M.; Patel, R. C.; Kaufman, P.; Joyce-Pruden, C.; O'Leary, N. *J. Chem. Educ.* **1993**, *70*, 342-344.

110. Hale, Penny S.; Maddox, Leone, M.; Shapter, Joe G.; Gooding, J. Justin. *J. Chem. Educ.* **2005**, *82*, 779-781.

111. Keating, Christine D.; Musick, Michael, D.; Keefe, Melinda H.; Natan, Michael J. *J. Chem. Educ.* **1999**, *76*, 949-955.

112. Craig, Vincent, S. J.; Jones, Anthony C.; Senden, Tim J. *J. Chem. Educ.* **2001**, *78*, 345-346.

113. González-Gaitano, Gustavo; Tardajos, Gloria. *J. Chem. Educ.* **2004**, *81*, 270-274.

114. Hernández-Benito; Jesùs; González-Mancebo, Samuel; Calle, Emilio; Garcia-Santos, M. Pilar; Casado, Julio. *J. Chem. Educ.* **1999**, *76*, 419-421.

115. Valero, M.; Rodriguez, L. J.; Velázquez, M. M. *J. Chem. Educ.* **1999**, *76*, 418-419.

116. Hernández-Benito; Jesùs; González-Mancebo, Samuel; Calle, Emilio; Garcia-Santos, M. Pilar; Casado, Julio. *J. Chem. Educ.* **1999**, *74*, 422-424.

117. Crane, Nicole, J.; Mayrhofer, Rudolph C.; Betts, Thomas A.; Baker, Gary A. *J. Chem. Educ.* **2002**, *79*, 1261-1263.

118. Van Hecke, Gerald R.; Karukstis, Kerry K.; Li, Hanhan; Hendargo, Hansford, C.; Cosand, Andrew J.; Fox, Marja M. *J. Chem. Educ.* **2005**, *82*, 1349-1354.

119. Demirbas, Erhan; Devonshire, Robin. *J. Chem. Educ.* **1996**, *73*, 586-589.

120. Dionisio, Madalena S. C.; Diogo, Herminio P.; Farinha, J. P. S.; Ramos, Joaquim J. Moura; *J. Chem. Educ.* **2005**, *82*, 1355-1360.

121. Chakraborty, Mriganka; Chowdhury; Devasish; Chattopadhyay, Arun. *J. Chem. Educ.* **2003**, *80*, 806-809.

122. Mougán, Manuel A.; Coello, Adela; Jover, Aida; Meijide, Franciso; Tato, José Vázquez. *J. Chem. Educ.* **1995**, *72*, 284-286.

123. Martin, Olga; Mendicuti, Francisco; Tarazona, Maria Pilar. *J. Chem. Educ.* **1998**, *75*, 1479-1481.

124. Andrews-Henry, Heather. *J. Chem. Educ.* **1994**, *71*, 357-358.

125. Pojman, John A.; West, William W.; Simmons, Jennifer. *J. Chem. Educ.* **1997**, *74*, 727-730.

126. Folmer, J. C. W.; Franzen, Stefan. *J. Chem. Educ.* **2003**, *80*, 813.

127. Barceló, A. Ros; Zapata, Z. M. *J. Chem. Educ.* **1996**, *73*, 1034-1035.

128. Coutinho, Ana; Prieto, Manuel. *J. Chem. Educ.* **1993**, *70*, 425-428.

129. Fancy, Nahyan; Mehl, Andrew F. *J. Chem. Educ.* **1999**, *76*, 646-648.

130. Richards, John L. *J. Chem. Educ.* **1993**, *70*, 685-689.

131. Williams, Kathryn R.; Adhyaru, Bhavin. *J. Chem. Educ.* **2004**, *81*, 1045-1047.

132. Williams, Kathryn R.; Tennant, Loretta H. *J. Chem. Educ.* **2001**, *78*, 349-351.

133. Crugeiras, Juan; Leis, J. Ramón; Rios, Ana. *J. Chem. Educ.* **2001**, *78*, 1538-1540.

134. Worley, John D. *J. Chem. Educ.* **1993**, *70*, 417-420.

135. Akers, Stephen M.; Conkle, Jeremy L.; Thomas, Stephanie N.; Rider, Keith B. *J. Chem. Educ.* **2006**, *83*, 260-262.

136. Myrick, M. L.; Greer, A. E.; Nieuwland, A.; Priore, R. J.; Scaffidi, J.; Andreatta, Danielle; Colavita, Paula. *J. Chem. Educ.* **2006**, *83*, 263-264.

137. Bowers, Peter G.; Rubin, Mordecai B.; Noyes, Richard M.; Andueza, Dagmar. *J. Chem. Educ.* **1997**, *74*, 1455-1458.

138. Lynam, Mary M.; Kilduff, James E.; Weber, Walter J, Jr. *J. Chem. Educ.* **1995**, *72*, 80-83.

139. Fan, Qunbai; Munro, Debra; Ng, L. M. *J. Chem. Educ.* **1995**, *72*, 842-845.

140. Mayer, Steven G.; Gach, Jeremy M.; Forbes, Evelyn R.; Reid, Philip J. *J. Chem. Educ.* **2001**, *78*, 241-242.

141. Hoffmann, Markus M.; Salter, Jason D. *J. Chem. Educ.* **2004**, *81*, 411-413.

142. Sorkhabi, Osman; Jackson, William M.; Daizadeh, Iraj. *J. Chem. Educ.* **1998**, *75*, 472-476.

143. Casarin, Carlos; Ibanez, Jorge. *J. Chem. Educ.* **1993**, *70*, 158-162.

144. Shabestary, Nahid; Khazaeli, Sadegh; Hickman, Richie. *J. Chem. Educ.* **1998**, *75*, 1470-1472.

145. Schulz, Pablo C.; Clausse, Danièle. *J. Chem. Educ.* **2003**, *80*, 1053.

146. Unpublished. Julio DePaula, Haverford College, Haverford, PA

147. Van Hecke, Gerald R.; Karukstis, Kerry K.; Haskell, Richard C; McFadden, Catherine S.; Wettack, F. Sheldon. *J. Chem. Educ.* **2002**, *79*, 837-844.

148. Gravelle, Steven; Langham, Beth; Geisbrecht, Brian. *J. Chem. Educ.* **2003**, *80*, 911-913.

149. Bengali, Ashfaq A.; Mooney, Kim E. *J. Chem. Educ.* **2003**, *80*, 1044-1047.

150. Ball, David B.; Miller, Randy M. *J. Chem. Educ.* **2004**, *81*, 121-125.

151. Diaz, Anthony; Radzewich, Catherine; Wicholas, Mark. J. Chem. Educ. 1995, 72, 937-938.

152. Casado, Julio; Izquierdo, Carmen; Fuentes, Santiago; Moyá, María Luisa. *J. Chem. Educ.* **1994**, *71*, 446-450.

153. Wickum, William G. *J. Chem. Educ.* **1998**, *75*, 1-3.
154. Ramachandran, B. R.; Halpern, Arthur M. *J. Chem. Educ.* **1999**, *76*, 1266-1268.
155. Sala, O.; Araki, K.; Noda, L. K. *J. Chem. Educ.* **1999**, *76*, 1269-1271.
156. Kaufman, M. J.; Trowbridge, C. G. *J. Chem. Educ.* **1999**, *76*, 88-89.
157. Khundkar, Lutfur R. *J. Chem. Educ.* **1996**, *73*, 1055-1056.
158. Cané, Elisabetta; Miani, Andrea; Trombetti, Agostino. *J. Chem. Educ.* **1999**, *76*, 1288-1290.
159. Pemberton, Jeanne E.; Wood, Laurie L.; Ghoman, Ghanshyam S. *J. Chem. Educ.* **1999**, *76*, 253-257.
160. Tubergen, Michael J.; Lavrich, Richard J.; McCargar, James W. *J. Chem. Educ.* **2000**, *77*, 1637-1639.
161. Craig, Norman C.; Lacuesta, Nanette N. *J. Chem. Educ.* **2004**, *81*, 1199-1205.
162. Parnis, J. Mark; Thompson, Matthew, G. K. *J. Chem. Educ.* **2004**, *81*, 1196-1198.
163. Dierenfeldt, Karl E. *J. Chem. Educ.* **1995**, *72*, 281-283.
164. Schor, H. H. R.; Teixeira, E. L. *J. Chem. Educ.* **1994**, *71*, 771-774.
165. Ganapathisubramanian, N. *J. Chem. Educ.* **1993**, *70*, 1035.
166. Ercolani, Gianfranco. *J. Chem. Educ.* **2005**, *82*, 1703-1708.
167. Ogren, Paul J. *J. Chem. Educ.* **2002**, *79*, 117-119.
168. Fetterolf, Monty L; Goldsmith, Jack G. *J. Chem. Educ.* **1999**, *76*, 1276-1277.
169. Volkov, Victor; Pfister, Rolf. *J. Chem. Educ.* **2005**, *82*, 1663-1666.
170. Pursell, Christopher J.; Doezema, Lambert. *J. Chem. Educ.* **1999**, *76*, 839-841.
171. Sabaté, Raimon; Freire, Llúcia; Estelrich, Joan. *J. Chem. Educ.* **2001**, *78*, 243-244.
172. Bahnick, Donald A. *J. Chem. Educ.* **1994**, *71*, 171-173.
173. Ramachandran, B. R.; Halpern, Arthur M. *J. Chem. Educ.* **1998**, *85*, 234-237.
174. Shalhoub, George M. *J. Chem. Educ.* **1997**, *74*, 1317-1319.
175. Anderson, Bruce D. *J. Chem. Educ.* **1997**, *74*, 985.
176. Horng, Miin-Liang; Quitevis, Edward L. *J. Chem. Educ.* **2000**, *77*, 637-639.
177. Delorenzi, Néstor J.; Araujo, César; Palazzolo, Gonzalo; Gatti, Carlos. *J. Chem. Educ.* **1999**, *76*, 1265-1266.
178. van Stam, Jan; Depaemelaere, Sigrid; Schryver, Frans C. *J. Chem. Educ.* **1998**, *75*, 93-98.
179. Clarke, Ronald J.; Oprysa, Anna. *J. Chem. Educ.* **2004**, *81*, 705-707.
180. Fery-Forgues, Suzanne; Lavabre, Dominique. *J. Chem. Educ.* **1999**, *76*, 1260-1264.

181. Shane, Edward C.; Price-Everett, Miranda; Hanson, Tonya. J. Chem. Educ. 2000, 77, 1617-1618.

182. Kvaran, Ágúst; Haraldsson, Árni Hr.; Sigfusson, Thorsteinn. J. Chem. Educ. 2000, 77, 1345-1347.

183. Prieto, M. Flor Rodríguez; Rodríguez, M. Carmen Rios; González, Manuel Mosquera; Rodríguez, Ana M. Rios; Fernández, Juan Carlos Mejuto. J. Chem. Educ. 1995, 72, 662-663

184. Jenkins, Amanda L.; Murray, George M. J. Chem. Educ. 1998, 75, 227.

185. Mays, Holger. J. Chem. Educ. 2000, 77, 72-76.

186. Roalstad, Shelly; Rue, Chad; LeMaster, Clifford B.; Lasko, Carol. J. Chem. Educ. 1997, 74, 853-854.

187. Jackson, Brad; Donato, Henry, Jr. J. Chem. Educ. 1993, 70, 780-782.

188. Fister, Julius C, III; Harris, Joel M.; Rank, Diana; Wacholtz, William. J. Chem. Educ. 1997, 74, 1208-1212.

189. Coppens, Paul. J. Chem. Educ. 2003, 80, 1316-1318.

190. Lorigan, Gary A.; Patterson, Brian M.; Sommer, Andre J.; Danielson, Neil D. J. Chem. Educ. 2002, 79, 1264-1266.

191. de Berg, Kevin C.; Chapman, Kenneth J. J. Chem. Educ. 2001, 78, 670-673.

192. Siders, Paul. J. Chem. Educ. 1997, 74, 235-236.

193. Muyskens, Mark A. J. Chem. Educ. 1997, 74, 850-852.

194. Mueller, Michael R.; McCorkle, Kent L.; J. Chem. Educ. 1994, 71, 169-170.

195. Wadsö, Lars; Smith, Allan L.; Shirazi, Hamid; Mulligan, S. Rose; Hofelich, Thomas. J. Chem. Educ. 2001, 78, 1080-1086.

196. Bryant, Philip A.; Morgan, Matthew E. J. Chem. Educ. 2004, 81, 113-115.

197. Crumpton, D. M.; Laitinen, R. A.; Smieja, J.; Cleary, D. A. J. Chem. Educ. 1996, 73, 590-591.

198. Merchán, M. Dolores; Salvador, Francisco. J. Chem. Educ. 1994, 71, 1085-1087.

199. Kennedy, A. P., Sr. J. Chem. Educ. 1997, 74, 1231-1232.

200. Ellison, Herbert R. J. Chem. Educ. 2005, 82, 1086-1088.

201. Iannone, Mark. J. Chem. Educ. 2006, 83, 97-98.

202. Knewstubb, P. F. J. Chem. Educ. 1995, 72, 261-263.

203. Marin-Puga, Gustavo; Guzman L., Miguel; Hevia, Francisco. J. Chem. Educ. 1995, 72, 91-92.

204. Burness, James H. J. Chem. Educ. 1996, 73, 967-970.

205. Williams, Kathryn R.; Collins, Sean E. J. Chem. Educ. 1994, 71, 617-620.

206. Lötz, A. J. Chem. Educ. 1996, 73, 195-196.

207. Nelson, Robert N. J. Chem. Educ. 1995, 72, 567-569.

208. Halpern, Arthur M. *J. Chem. Educ.* **2002**, *79*, 214-216.
209. Victoria, L.; Arenas, A.; Molina, C. *J. Chem. Educ.* **2004**, *81*, 1333-1336.
210. Digilov, Rafael M. *J. Chem. Educ.* **2002**, *79*, 353-355.
211. Seckin, Turgay; Kormali, Suphi M. *J. Chem. Educ.* **1996**, *73*, 193-194.
212. Zhang, Xin; Lui, Shouxin; Li, Baoxin; An, Na; Zhang, Fan. *J. Chem. Educ.* **2004**, *81*, 850-853.
213. Baonza, Valentín García; Cáceres, Mercedes; Núñez, Javier. *J. Chem. Educ.* **1996**, *73*, 690-693.
214. Heavers, Richard M. *J. Chem. Educ.* **1997**, *74*, 965-967.
215. Kugel, Roger W. *J. Chem. Educ.* **1998**, *75*, 1125-1129.
216. Blanco, Luis H.; Romero, Carmen M.; Munar, Ricardo. *J. Chem. Educ.* **1995**, *72*, 1144-1146.
217. Smith, Christopher W.; Cooke, Jason B.; Gliski, Robert J. *J. Chem. Educ.* **1999**, *76*, 227-228.
218. Erickson, Luther E.; Morris, Kevin. *J. Chem. Educ.* **1996**, *73*, 971-974.
219. Hansen, Keith C.; Zhou, Zhou; Yaws, Carl L.; Aminabhavi, Tejraj M. *J. Chem. Educ.* **1995**, *72*, 93-96.
220. Kundell, Frederick A. *J. Chem. Educ.* **1999**, *76*, 542.
221. Kabza, Konrad; Gestwicki, Jason E.; McGrath, Jessica L. *J. Chem. Educ.* **2000**, *77*, 63-65.
222. Cobb, C. L.; Love, G. A. *J. Chem. Educ.* **1998**, *75*, 90-92.
223. Clay, J. T.; Walters, E. A.; Brabson, G. D. *J. Chem. Educ.* **1995**, *72*, 665-667.
224. Wilczek-Vera, Grazyna. *J. Chem. Educ.* **1995**, *72*, 472-475.
225. Aroti, Andria; Leontidis, Epameinondas. *J. Chem. Educ.* **2001**, *78*, 786-788.
226. Huang, Xirong; Yang, Jinghe; Zhang, Wenjuan; Zhang, Zhenyu; An, Zeshong. *J. Chem. Educ.* **1999**, *76*, 93-94.
227. Hwang, Jimmy S.; Oweimreen, Ghassan A. *J. Chem. Educ.* **2003**, *80*, 1051-1052.
228. Byrne, John P. *J. Chem. Educ.* **1994**, *71*, 531-533.
229. Burley, Joel D.; Roberts, Alisa M. *J. Chem. Educ.* **2000**, *77*, 1210-1212.
230. Marcotte, Ronald E.; Wilson, Lenore D. *J. Chem. Educ.* **2001**, *78*, 799-800.
231. El Seoud, Omar A.; Takashima, Keiko. *J. Chem. Educ.* **1998**, *75*, 1625-1627.
232. González, Gabriel; Martinez, Manuel. *J. Chem. Educ.* **2005**, *82*, 1671-1673.
233. Mowry, Sarah; Ogren, Paul J. *J. Chem. Educ.* **1999**, *76*, 970-973.
234. Williams, Kathryn R.; Adhyaru, Bhavin; Timofeev, Julia; Blankenship, Michael Keith. *J. Chem. Educ.* **2005**, *82*, 924-925.

235. Buccigross, Jeanne M.; Metz, Christa; Elliot, Lori; Becker, Pamela; Earley, Angela S.; Hayes, Jerry W.; Novak, Michael; Underwood, Gayl A. *J. Chem. Educ.* **1996**, *73*, 364-367.

236. Ramachandran, B. R.; Halpern, Arthur M. *J. Chem. Educ.* **1996**, *73*, 686-689.

237. Birk, James P.; Walters, David L. *J. Chem. Educ.* **1993**, *70*, 587-589.

238. Rosbrugh, Ian M.; Nishimura, S.Y.; Nishimura, A. M. *J. Chem. Educ.* **2000**, *77*, 1047-1049.

239. Ramachandran, B. R.; Halpern, Arthur M. *J. Chem. Educ.* **1997**, *74*, 975-978.

240. Hemalatha, M. R. K.; NoorBatcha, I. *J. Chem. Educ.* **1997**, *74*, 972-974.

241. Nóbrega, Joaquim; Rocha, Fábio R. P. *J. Chem. Educ.* **1997**, *74*, 560-562.

242. Bigger, Stephan; Watkins, Peter J.; Verity, Bruce. *J. Chem. Educ.* **2003**, *80*, 1191-1193.

243. Dakkouri, M.; Bodenmüller, W. *J. Chem. Educ.* **1997**, *74*, 556-559.

244. Jonnalagadda, S. B.; Gollapalli, N. R. *J. Chem. Educ.* **2000**, *77*, 506-509.

245. Cugeiras; Juan; Leis, J. Ramón; Ríos, Ana. *J. Chem. Educ.* **2001**, *78*, 1538-1540.

246. Williams, Kathryn R. *J. Chem. Educ.* **2000**, *77*, 626-628.

247. El Seoud, Omar A.; Bazito, Reinaldo C.; Sumodjo, Paulo T. *J. Chem. Educ.* **1997**, *74*, 562-565.

248. Churio, M. S.; Grela, M. A. *J. Chem. Educ.* **1997**, *74*, 436-438.

249. Wade, E. A.; Clemes, T. P.; Singmaster, K. A. *J. Chem. Educ.* **2000**, *77*, 898-900.

250. Peral, José; Trillas, María; Domènich, Xavier. *J. Chem. Educ.* **1995**, *72*, 565-566.

251. Willett, Kristine L.; Hites, Ronald A. *J. Chem. Educ.* **2000**, *77*, 900-902.

252. Prypsztejn, Hernán; Negri, R. Martín. *J. Chem. Educ.* **2001**, *78*, 645-648.

253. Maestri, Mauro; Ballardini, Roberto; Pina, Ferdando; Melo, Maria João. *J. Chem. Educ.* **1997**, *74*, 1314-1316.

254. Raymundo-Piñero, E. Raymundo; Cazorla-Amorós, D.; Morallón, E. *J. Chem. Educ.* **1999**, *75*, 958-961.

255. Copper, Christine L.; Koubek, Edward. *J. Chem. Educ.* **1998**, *75*, 87-89.

256. Fischer, Jonathan D.; Whitten, James E. *J. Chem. Educ.* **2003**, *80*, 1451-1454.

257. Pojman, John A.; Craven, Richard; Leard, Danna C. *J. Chem. Educ.* **1994**, *71*, 84-90.

258. Strizhak, Peter; Menzinger, Michael. *J. Chem. Educ.* **1996**, *73*, 868-873
259. Strizhak, Peter; Menzinger, Michael. *J. Chem. Educ.* **1996**, *73*, 865-868.
260. Clark, Charles R. *J. Chem. Educ.* **1997**, *74*, 1214-1217.
261. Drok, K. J.; Ritchie, I. M.; Power, G. P. *J. Chem. Educ.* **1998**, *75*, 1145.
262. Hodgson, Steven C.; Ngeh, Lawrence N.; Orbell, John N.; Bigger, Stephen W. *J. Chem. Educ.* **1998**, *75*, 1150-1153.

Problem-Solving Issues
in Quantum Mechanics

Chapter 9

Existence of a Problem-Solving Mindset among Students Taking Quantum Mechanics and Its Implications

David E. Gardner and George M. Bodner

Department of Chemistry, Purdue University, West Lafayette, IN 47907

This chapter summarizes the results of a qualitative research study of undergraduate chemistry and physics students enrolled in introductory quantum mechanics courses. We found that many of the problems the students encountered when learning quantum mechanics were not the result of a misunderstanding of the concepts being taught but the result of their employing non-productive strategies while studying and doing the homework. The goal of this chapter is to describe the problem-solving mindset the students brought to the learning of quantum mechanics as a basis for thinking about changes in the way quantum mechanics, in specific, and physical chemistry, in general, are taught.

Introduction

Several years ago, *The American Journal of Physics* published an issue devoted to the teaching and learning of quantum mechanics with the expressly stated hope that the articles "will help people enhance their teaching" (*1*). In describing the articles contained in this issue, the editors stated:

> Some of the articles address the difficulties students have learning particular aspects of quantum mechanics. Others describe different interpretations, formulations, and representations in quantum mechanics. Still others discuss novel applications or some of the more subtle conceptual issues in quantum mechanics. A few of the articles address the integration of workable and affordable quantum mechanics experiments into the undergraduate curriculum.

While the authors of these papers were typically motivated to improve quantum mechanics by clarifying some particular aspect or by making the subject more palatable, their recommendations were seldom supported by any research on the teaching and learning of quantum mechanics. The trend toward research on the educational aspects of advanced topics such as quantum mechanics is a recent one (*2-6*).

Johnston and coworkers (*5*) found that students enrolled in quantum mechanics classes did not have coherent or internally consistent conceptual frameworks; moreover, the understanding they did possess seemed to be fragmented and isolated. They found little evidence that the students had a "deep" understanding, suggesting instead that students' understanding was superficial. The authors found this particularly troubling because the students used in the study were "good" in every criteria commonly used in a university. They noted that common methods of instruction and assessment emphasize the importance of "facts" rather than the mental structure in which those facts are embedded. Thus, traditional instruction is "encouraging exactly the kind of fragmented conceptual development being observed" (p. 442). They also noted that their data showed no evidence that students' mental frameworks improve over time. They informally presented the same survey to both second-year students and graduate students and found few differences in students' mental models of these phenomena.

Context of this Study

This chapter describes the problem-solving mindset that many chemistry students bring to quantum mechanics classes that was discovered during a study of the lived experiences of students struggling to learn quantum mechanics. The purpose of this chapter is two-fold. First, by describing this problem-solving mindset, we wish to create a lens through which we, as educators, may critically examine our students. Although many of the features of the problem-solving mindset are not new and have been discussed in other contexts (7-9), it is hoped that this conceptualization will lead to new insights. Second, accepting that students have a problem-solving mindset leads to a variety of implications for possible changes in the way quantum mechanics, in specific, and physical chemistry, in general, are taught.

Our understanding of the problem-solving mindset evolved out of qualitative research that examined students' difficulties in learning quantum mechanics. Initially, we were primarily interested in identifying the conceptual challenges students had to overcome in learning quantum mechanics, while also examining the experiences they encountered while engaged in these challenges. To accomplish this, we examined the actions and behaviors of students while they were in class, while they were working on homework, and while they were studying. We also examined the students' attitudes toward these activities and their ideas about what they thought they should be doing. Early indications from the data, however, suggested that many of the problems the students encountered when learning quantum mechanics were less the result of a misunderstanding of the concepts being taught and more the result of employing non-productive strategies while studying and doing the homework. Thus, an additional focus was developed that examined the approaches students used to learn quantum mechanics. It was this final focus that led to the discovery of the problem-solving mindset.

Definitions of Terms

For some time, we have been trying to differentiate between two terms that are often used more or less synonymously by chemists: *exercise* and *problem*. Fifteen years ago, we differentiated between these terms as follows:

> We all routinely encounter questions or tasks for which we don't know the answer, but we feel confident that we know how to obtain the answer. When this happens, when we know the sequence of steps needed to cross the gap between where we are and where we want to be, we are faced with an exercise not a problem. (7, p.21)

We noted that "... there is no innate characteristic of a task that inevitably makes it a problem. Status as a problem is a subtle interaction between the task and the individual struggling to find an appropriate answer or solution." We went on to argue that it is the existence of a well-defined algorithm, constructed from prior experience, which turns a question into an exercise.

Over the years, we have tried to emphasize the difference between exercises and problems by referring to the first as a *routine exercise* and the second as a *novel problem* (*10*). To illustrate this difference, we have frequently quoted a definition of problem solving introduced by Wheatley more than 20 years ago: "Problem solving is what you do when you don't know what to do" (*11*).

In the course of this chapter, we will routinely use the terms *problem solving* and *problem-solving mindset*. We do this deliberately because we believe that the tasks students are working on in quantum mechanics are, in fact, problems for them at the time that they first encounter them.

Background

The participants in this study were chemistry, chemical engineering, or physics students in various upper-level classes at Purdue University in which quantum mechanics represents a sizeable portion of the material covered, if not the focus of the entire course. We had prior approval from the instructors to observe their classes and to interact with their students, and from the institution's IRB to do the study. All of the students who participated in the study were volunteers. Students were informed that participation in the study was not a criterion their instructors would use when assigning their grades. The students who participated in this study represented a good cross-section of the population enrolled in these classes, from students who did very well to those who struggled to pass.

As noted earlier, we wanted to understand the students' overall general experience of quantum mechanics, as well as the difficulties they encountered learning the material. There can be a wide variety of reasons, for example, why a student might say: "I'm lost." Our goal was to interpret what that student meant, be it a simple confusion about the notation used in class or a fundamental misunderstanding of some important physical concept. Accurately interpreting the words and actions of the students required close and familiar knowledge of the students within the context of physical chemistry and quantum mechanics. Thus, the research was based on interacting with students in order to see what they did and how they did it; as well as listening to what they said, what they did not say, and how they expressed themselves with regard to quantum mechanics.

Methods of Data Collection

The first source of data for this study was a set of field notes based on classroom observations, a logical first step in a study of students' experiences in learning physical chemistry. For each of the courses from which data were collected, the first author attended the lectures for the portion of the class that covered quantum mechanics. Any thoughts, comments, ideas, or questions were written into a notebook in the form of field notes. Classroom observations focused on: (1) watching and listening to what the professor was presenting and how it was presented; (2) non-verbal student responses to instruction (sleeping, attentiveness, and so on); and (3) verbal student responses such as questions and comments. Of these, questions and comments were the most valuable because they provided a window into what the students were thinking.

A second source of data was 3 x 5 cards that were distributed to the students at the beginning of each class. We asked the students to write down any questions or comments regarding the day's lecture and to hand the cards to the researcher, not the instructor. The students were told that their names were not required and that their comments would be passed along anonymously to the instructor. In all, over 300 cards were collected. We estimated that between one-third and one-half of the students handed in one or more cards. Of these students, there were ten to fifteen who were very prolific, and wrote the majority of the cards.

When combined with questions asked aloud during lecture, the 3 x 5 cards provided clues as to both what and how students were thinking about quantum mechanics. The cards also allowed the students to express their feelings and impressions of class while the class was happening, when these thoughts and feelings were still fresh in their mind. This provided direct insight into their experiences of a class in quantum mechanics. The cards also provided a mechanism for validating classroom observations that focused on the difficulties students seemed to be experiencing during lecture.

The third and most important source of data was tutoring sessions. At the beginning of the semester, we offered free tutoring for the students in exchange for participating in the study. Students usually came for help on a homework assignment or while preparing for an exam. Sessions were usually small, one or two people, although on several occasions we hosted groups of six to ten students. In all, thirty-seven different students came in for help. Twenty-two of these sessions were audiotaped and were later transcribed for analysis.

These tutor-sessions/interviews were very loosely organized, with the direction of the discussion being guided by the students' questions. This is different from traditional research interviews, where the researcher is the one asking the questions. However, when appropriate, the researcher did ask the students questions related to various aspects of the class and their understanding of the material. Marton (12) points out that the interviews in some of his phenomonographical research were almost like pedagogical situations. The tutor sessions in this study crossed that line because they were, in fact, pedagogical situations.

The final data source was a set of traditional interviews conducted at the end of the semester. Twelve students participated in these interviews, four from class B (described below) and eight from class C2. The interviews were loosely based on an interview guide and ran between 30 and 45 minutes in length. The purpose of the interviews was to inquire more deeply and directly into the experiences the students had during the semester.

Data were collected from students enrolled in three different courses. Class A was a one-semester introductory quantum mechanics course intended for junior physics majors that typically enrolled about 10 students. Class B was the second-half of a two-semester physical chemistry course for chemistry majors that typically enrolls 30-40 students. The first semester of this course focuses primarily on thermodynamics; the second-half spends the first two-thirds of the semester on quantum mechanics and then concludes with a discussion of statistical mechanics. Class C is offered every semester for junior-year chemical engineering majors, and was observed three times: C1, C2, and C3. C1 and C3 were offered during the fall semester, when the mainline population of chemical engineering majors take the course and had enrollments of approximately 70 students. C2 was offered in the spring semester and is frequently taken by students who have done a "co-op" or internship in industry, which requires them to be off-campus for a semester at a time. C2 had an enrollment of around 30 students. The material in Class C is quite similar to the material offered in Class B. The first three-quarters of this class covers quantum mechanics, the remaining time is spent on statistical mechanics.

Frameworks for the Study

The theoretical framework for this study was the constructivist theory of knowledge, which holds that knowledge is created in the mind of the learner (13, 14). The methodological framework was hermeneutical phenomenography.

Hermeneutics is a field of study concerned with the interpretation of texts, either written or spoken (15). Within the context of educational research it has been defined as the process of providing a voice to individuals or groups of individuals who either cannot speak for themselves or are traditionally ignored (16). Patton (17) states that "to make sense and interpret a text, it is important to know what the author wanted to communicate, to understand intended meanings, and to place documents in a historical and cultural context."

Phenomenography is the study of lived experiences (18). According to Marton (19), the goal of phenomenography is a "description, analysis, and understanding of experiences." To achieve this goal, a researcher catalogues and describes the various conceptions and perceptions of a phenomenon, as well as looks for the underlying meanings and connections between those conceptions. Through this process, the research turns something that is "unthematized" into an object of focal awareness (*12*). As awareness is increased, the object or idea is brought from the subconscious to the conscious, where it can be overtly talked about and discussed because, by definition, as long as the experiences remain implicit, they cannot be discussed. The effect of this process is to empower the subjects involved in the study by "giving them voice" or "providing language" for them to think about and discuss their experiences in ways that they were previously unable to do.

Data Analysis

The initial step in data analysis was transcription of the raw data present in the audiotapes into more easily handled text. During the early phase of data analysis, which was concurrent with much of the data collection, we tried to discern trends or patterns in the data. In addition, we reflected on the data and tried to develop theories or explanations to account for our observations. This process became more formalized in the later stages of data analysis, which occurred after the majority of the data had been collected. For the later phases, we coded the data following the guidelines set forth by Huberman and Miles (*20*) using the Atlas.ti (*21*) software package to help with data management. The encoded data allowed the trends and patterns in the data to emerge in a more refined manner than through our preliminary analysis.

The Problem-Solving Mindset

The approach that the majority of the students in this study took to learning quantum mechanics was based on a single, common, unstated assumption: their goal was to solve problems. This assumption was so pervasive it can be best termed a "problem-solving mindset." Students with this mindset organize their behavior and thinking around the idea that the main objective in this class, as in so many other courses they have taken, was to solve problems. Moreover, students with this mindset perceive that the rationale for taking the course was to learn additional ways of solving new and more comprehensive types of problems. The problem-solving mindset had an effect on two major aspects of the students' experiences in class: the expectations they had of what the class *should* be like and the behaviors the students used to get through class.

Expectations

The students in this study were primarily juniors and seniors and had considerable experience in science and math courses prior to studying quantum mechanics. For the most part, they expected that the class would focus on problem solving. These expectations form the core of the problem-solving mindset. It is reasonable to assume that the expectations they brought with them were based on their prior experiences in science classes. Although these expectations were manifested in a variety of ways, the three most common expectations were:

- The students expected to see numerical examples.
- The students expected answers to be precise and correct.
- The students expected the material to be useful.

A dominant theme in the data was the expectation that the students would be provided with numerical examples. A sample of student responses that address this issue, which were all collected within two days of each other during the first month of the semester, are given below.

> "Can we get a numeric example done in class to calculate E(T,V)?" [Card response, class C2]

> "Why are we learning stat. thermo? What are we going to use it for? A numeric example would be good." [Card response, class C_2]

> "Schrödinger Equation: I have to see a numerical example." [Card response, class B]

> "Could you give an actual example of a Ψ?" [Card response, class B]

The underlying assumption in several of these cards was that a numerical example would overcome all of the troubles the student was having. The critical feature in comments that reflected the first of the students' expectations was that examples were only valid if they were numerical. When instructors presented non-numerical examples, the students generally viewed the information as more "theory." Consider Craig's reaction to the question of whether discussion of an example based on the particle in a box was what he wanted to see.

> I: ... people kept saying "Oh, I want to see examples. We want to [see] examples." And finally he showed you an example of the particle-in-a-box. And ... did you like that example of a particle in a box? Doesn't that make you happy as an example?

> Craig: I don't really think that's an example. When am I ever really going to deal with a particle in a box?

I: Well, well ... okay. One of ...

Craig: I understand that's an example, but that wasn't the example I was ... needing to see.

Craig made several interesting points in this quote from an interview at the end of the semester. The first was that the particle-in-a-box was not a good example because it wasn't useful; as Craig pointed out, it is unlikely that he would ever deal with the particular case of a one-dimensional particle in a square well. Second, and even more intriguing, Craig commented that although he understood that the particle-in-a-box was an example, it was not the one he needed to see.

Although it is not explicitly stated in the above quote, it was clear from Craig's interview that what he wanted were numerical examples. Other students made their preferences known in their responses during tutoring sessions. In a discussion of the differences between general chemistry and physical chemistry courses, the interviewer talked about problem-solving questions that involved calculating the density of a sample from measurements of the mass and volume of the sample. He then noted:

I: ... and often times, at first you guys ask — say, "we want examples." Have any of you ever thought this during class?

Gunther: Yes! [There is also general agreement of the group]

I: The examples that you want, do you want numbers in your examples? Or do you want like the examples that he shows you?

Group: Numbers!

Gunther: Much prefer numbers. [laughing]

One aspect of this conversation that cannot be captured in a transcript was the enthusiasm and excitement the students had during this exchange. The interviewer's question seemed to strike a chord with this group of students because it tapped into some of the frustration that they experienced in the class that revolved around the difference between what they had come to expect in a chemistry course and what was happening in their quantum mechanics course.

The second expectation that students with a problem-solving mindset had related to the answers to questions they were asked on homework and exams. These students expected that the answers they obtained as the result of their calculations were the "true" and "correct" values. In other words, the equation used during a problem was expected to give a value that exactly corresponded to the real world. The students did not acknowledge that an equation that was correctly applied and solved might not give the correct real world value. Elsewhere we have argued that this is a consequence of the approach to instruction that characterizes so many science and mathematics classrooms (*22*). Discussions we have had with many physical chemistry instructors suggest that they often begin their discussion of thermodynamics by talking about equations of state, such as the ideal gas law. They then introduce the van der Waals equation as an alternative equation of state and often compare the predictions of these two equations for a sample of a real gas at a given temperature confined in different containers of ever-decreasing volume. Not one of them, so far, has admitted taking the next logical step — comparing the results of both calculations with experimental data.

This expectation that the results of a calculation that has been done "correctly" will themselves be "correct" also extends to the method by which the problem was solved. Unfortunately, because few equations in quantum mechanics can be solved exactly, scientists are forced to use numerical computational methods to calculate values. Strictly speaking, even though it is possible to calculate these values to arbitrary levels of precision, such solutions are only approximations of the "exact" value. Because of the difficulty involved in most of the calculations, a number of students expressed frustration with the notion that all of their hard work had only yielded them an approximate answer.

The third expectation exhibited by students with a problem-solving mindset is the desire to learn useful material. While such a desire is not unusual, what was surprising was how the students defined the term "useful." For many of the students in this study, usefulness was equated with the ability to solve problems, as shown in the following quotes:

"What can we use these equations for? (i.e. what physical use is it)" [Card response, class B]

"I have no idea why it was important that we learn how to derive the wave equation. Why didn't we just learn what it is and what it's used for." [Card response, class C3]

"Major thought in my head all lecture: 'So What? What does all this math do for me? What do people use this stuff for?'" [Card response, class C3]

"I just think that there should be more chemistry applied and not so much derivation information. Tell us more how it applies to chemistry." [Card response, class C3]

"My question deals with the overall picture, I do not understand what this is leading to. ... what is physical chemistry used for? I have had physics and thermodynamics, and by the lectures I have learned about the Boltzmann distribution, but I guess I just don't understand why I care to use these things? Can you explain where these applications are useful? Theory is great, but I like to apply. Thanks! [continued on the back of card] I just wanted to say, I know p-chem has applications, just the applications for what we are learning." [Card response, class C2]

There were two distinct, although related aspects, to this focus on solving problems that revolved around the terms *what* and *how*. The first concerned the applications of the material, i.e. given these concepts and tools, what problems can I solve? The second concerned how that material was used, i.e. given these concepts and tools, how are these problems solved? The students needed to understand both the "what" and the "how" before they were willing to accept that learning the material presented in class was useful.

Behaviors

Because the problem-solving mindset shaped what students expected the class would be like, it also influenced their decisions on what they needed to do to be successful in class. As noted above, students with a problem-solving mindset believed that the purpose of class was to solve problems. For these students, their job was to find the answers to the problems the instructor presented. Moreover, they expected that by looking through the textbook or their class notes they would either find the answer or a solved example problem exactly like the one being asked.

In order to do what they thought would make them successful in the course, students with a problem-solving mindset adopted a number of strategies that they felt made obtaining the answer easier. These strategies were superficial because the students concentrated on shallow, surface-level features of the tasks they encountered, not on any deep conceptual understanding. Some of the strategies students used were:

- Scanning the book for equations with visual, surface-level similarity.
- Mimicking the solutions found in the textbook or notes.
- Working backwards from the answers in the back of the book.

In spite of adopting these superficial strategies, the students were able to perform well in the course even though they might have possessed little or no conceptual understanding of the material.

A good example of the use of superficial strategies can be found in the following extract from a tutoring session.

> I: Okay. Do you know what they are doing here? Rather than just following exactly what they do in the book.
>
> Amanda: uh-uh [negative].
>
> I: Not a clue?
>
> Charlie: Nope, not a clue.
>
> Bob: Concept-wise, I don't understand anything that we are doing. I am just using my math skills ... to do what they tell me to do. They say, show that this equals this. So, I am using the skills I know from math and calculus and I'm trying to... [several lines of text omitted] ... yeah, but I don't know what the heck it means. I don't know what the heck the values in there are telling me or anything.
>
> I: Okay.
>
> Charlie: Because, the end of the book ... the answers don't shine the light on anything.
>
> Amanda: And, plus, it's not necessary. I mean, he does not make it necessary for us to understand.
>
> Charlie: Yeah.

These students admitted that they did not understand what they were doing; they were only mimicking the solution to a similar problem in the book. Charlie indicated that they had also attempted checking with the answers at the back of the book with little success. As Bob commented, because he did not understand the concepts, his strategy was to get by using his math skills, which were reasonably good.

It was clear from both their tone and the rest of the conversation that these students expended a considerable amount of effort in trying to understand the material, and that they had little success with anything that they tried. On the audiotape, the frustration in their voices is clear. The most shocking comment is Amanda's when she stated that it was not necessary to understand the material. She realized that they would be able to get through the course based only on their ability to solve problems, and, in essence, fake their way through the class because the methods of assessment did not actually measure their conceptual understanding. Although superficial strategies are often very useful, over-reliance on them can be problematic because students who are adept at using these strategies may mask shortcomings in conceptual understanding, even to the point where they may do quite well in class based solely on their problem-solving skills.

In many ways, what we saw was a demonstration of Herron's (*14*) principle of least cognitive effort. In general, students will do the least amount of work they can get away with and still get the grade that they want. Amanda and her friends tried to understand and were unsuccessful. Once they realized that it did not matter, and started to adopt the attitude of looking for the path of least resistance, they tended to adopt these superficial strategies with even more vigor.

Skemp (*23*) differentiated between relational and instrumental learning. Students who do not focus on the conceptual aspect of the task before them are instrumental learners; they focus on the necessary rules and formulas needed to produce correct answers during assessment. Instrumental learning, such as mimicking a solution in the book, is a very different activity than studying that solution with the intent of being able to better understand the problem, which Skemp describes as relational learning, i.e., learning with the intent to develop conceptual understanding.

The effect of the problem-solving mindset was clear: students made few connections with the conceptual aspects of the material. The evidence for this was found in how these students studied for their exams. They tended to perceive that there were many equations they had to learn, and, as the following extracts from tutoring sessions illustrates, they adopted the technique of brute force memorization.

> Mary: And I don't understand how we can apply any of this, really. What am I supposed to do with this stuff? On an exam is there going to be like, derive this equation from something else? Or am I supposed to know how to use these five million equations that I don't know.

> Larry: I'm not sure what the purpose of this ... of what he has done so far, except to memorize a bunch of facts and then spit them out at exam time. ... So what's the point of memorizing a bunch of facts if I 'm not ... if they are wrong anyway? Sure, I can memorize some facts about something that I am never going to use. To broaden the knowledge, or expand the horizon so to speak. [laughs].

While this was not memorization in the sense of being able to correctly reproduce the equations from memory, it was memorization in the sense of rote or non-meaningful learning (*24*), because the students did not really understand the physical significance to the equations.

Discussion

The implications of a problem-solving mindset go beyond students' approach to learning quantum mechanics and their success, or lack thereof, in physical chemistry courses. This mindset is a reflection of the students' understanding of the nature of science. Through their expectations that answers to problems should be both precise and correct in the sense of matching experimental results, students operating with a problem-solving mindset demonstrate a belief in the absolute nature of scientific knowledge. These students often fail to recognize that what they are learning are models of physical phenomena and the term *model* is best used in the sense of the following definition from the Oxford English Dictionary: "A simplified or idealized description or conception of a particular system, situation, or process, often in mathematical terms, that is put forward as a basis for theoretical or empirical understanding, or for calculations, predictions, etc.; a conceptual or mental representation of something." Moreover, the models that are presented in the junior-level physical chemistry class are often the simplest ones because these are the models that are the easiest to understand and manipulate, not because they give the best results.

The problem-solving mindset is not compatible with the actuality of science, which is that much scientific knowledge represents not the absolute, final truth, but our current best understanding. As a result, this mindset can impede conceptual learning. Consider, for example, the effect of asking a student to compare two competing theories or models. Because the focus of this comparison in the mind of the student is on obtaining the correct numerical answer, when two theories give quantitatively different answers to a problem, a student with a problem-solving mindset will conclude that one of the theories must be wrong. Even though there may be solid pedagogical reasons for studying both models, this student will likely expend little effort on learning the "bad" model because it gives "wrong" answers which are therefore considered useless and of little value.

A second issue that may arise for students operating with a problem-solving mindset is a failure to recognize the creative nature of science. From a problem-solving mindset, science is a linear march from an equation and a set of initial conditions toward a single, unambiguous final answer. Such a journey requires little creativity and discounts the reality that the practice of science, in general, involves the creation of models to explain and then predict phenomena. Indeed, the idea that modeling plays an important role in science is incongruous with a problem-solving mindset.

Compare the process a scientist uses in building a mathematical, theoretical model to the process a student with a problem-solving mindset uses to solve a typical textbook problem. When creating a model, the scientist first identifies the relevant aspects of a phenomenon and then generates a mathematical description encapsulating those aspects. For the student, the process is reversed;

he or she typically starts with an equation and then connects the symbols it represents to the phenomenon. For the practicing scientist, the building of a model is a cyclical process through which the model is refined and becomes progressively better. From the perspective of the student, multiple cycles through the problem are to be avoided whenever possible. For the student, proficiency is demonstrated by being able to move directly through the problem without being sidetracked. For the scientist, evaluation of the model is based on the needs of the scientist creating the model. For the student with a problem-solving mindset the authority for judging whether an answer is correct resides with an external authority, either the answers at the back of the book or the instructor.

Our results raise several unanswered questions. First, and perhaps foremost, how widespread is the problem-solving mindset? Although the problem-solving mindset was identified based on analysis of comments made by the students involved in this study, it is not reasonable to assume that these students are unusual or unique. There is every reason to believe that these students are similar to their peers, upper-level chemistry, physics and chemical engineering students throughout the country.

Is the problem-solving mindset found in less advanced students, i.e., those in general chemistry? Probably. There were no indications in our data that the problem-solving mindset had replaced any previous conception of science. Nor are there any reasons to believe that anything in the standard curriculum is likely to have produced any major changes in students' mindset prior to enrolling in the junior-level physical chemistry classes used in this study. Moreover, based on personal experiences in teaching general chemistry, there are abundant indications that the problem-solving mindset is present there too.

Having identified the problem-solving mindset in our study, are our results consistent with prior work? We believe the answer is: Yes. In their summary of research in physics education, Redish and Steinberg (25) state that even excellent students in introductory physics use problem-solving techniques characterized as "dominated by superficial mathematical manipulations without deeper understanding" (p. 25). Moreover, Redish and Steinberg claim many introductory physics students treat physics as a collection of isolated facts, choosing to focus on memorizing and using formulas instead of learning the underlying concepts. Furthermore they cite the work of Hammer (26) who found students' approach to physics problems was counterproductive to helping them develop a strong conceptual understanding.

Results similar to those discussed by Redish and Steinberg (25) have also been found in chemistry. Carter (27) found that general chemistry students' beliefs about the nature of chemistry affected their ability to solve problems and learn chemistry. She noted that instrumental learners view chemical knowledge as a series of rules and facts to be memorized. Moreover, such students made few, if any, connections between these facts. They believed their job was to reproduce the pieces of knowledge presented to them and considered assigned problems to be opportunities to regurgitate that knowledge, not opportunities to develop better conceptual understanding.

What effect does a problem-solving mindset have on students? For the students in this study, most of the difficulties they had learning quantum mechanics resulted from the incongruity between the structure of quantum mechanics and the problem-solving mindset, and not from conceptual difficulties within the material of quantum mechanics.

In order to anticipate the effect of a problem-solving mindset on classes other than physical chemistry, it is useful to consider the genesis of the mindset and ask the question: Why did so many of the students in this study have a problem-solving mindset? A possible explanation is conditioning. As mentioned above, the students in this study had years of experience in science and math classes before they came to physical chemistry and most of these classes, both at the university level and before, were organized around a central theme of solving problems and exercises.

Consider the experiences that many instructors provide students in a typical general chemistry class. First, they are assigned many types of problems to solve. Moreover, their instructors often tell them that one of the best ways to learn the material and study for the exams is to work lots of problems. Next, as part of instruction, they are provided with the necessary set of rules and algorithms needed to solve those problems efficiently. When they are introduced to concepts, such as density or equilibrium constants, the lecturer tends to go over an example. With few exceptions, such examples consist of numerical values being plugged into the equation to yield a single numerical answer. These students are then assessed on their ability to solve such problems — perhaps because it so much easier to create and grade examinations for which the question have a single correct numerical answer — which further emphasizes the importance of problem solving. If their other educational experiences bear much resemblance to the experiences in a typical general chemistry class, we should not be surprised that they develop a problem-solving mindset.

Skemp (23) provides a theoretical explanation for such behavior in his theory of learning and education. As noted previously, Skemp distinguishes between relational learning, which is focused on the understanding of concepts and the development of schema, and instrumental learning, which tends to focus on learning the necessary rules for finding the right answers needed to make the grade. Memorizing a rule is quicker than investing the effort needed to develop the schema required for conceptual understanding. As Skemp points out, the paradox is that although more effort is required to learn a schema, there is less to remember once the schema is learned because, once learned, "an indefinitely large number of particular plans can be derived" (p. 260).

For classes other than physical chemistry, it is quite likely that the problem-solving mindset would have little adverse effect on students' performance in terms of the grades they earn in the course. Therein lies the problem. If our proposed explanation of the origins of this mindset is correct, then the problem-solving mindset is what makes students seem to be successful in many classes. Thus, the mindset which develops because the students want to achieve good grades may actually hinder conceptual understanding and, in addition, reinforce misconceptions about the nature of science.

The final and most important question is: What effect does the problem-solving mindset have on instruction? The goal of this chapter is not to condemn problem-solving activities. Problem solving is a valuable pedagogical tool and having our students develop good problem-solving skills is a laudable educational outcome. It is our belief that the issue is not problem solving itself, but on the over-reliance on problem solving in instruction and assessment that leads our students into adopting a problem-solving mindset. In order to challenge the problem-solving mindset, the logical conclusion is that we need to provide students other ways of experiencing science instruction and, more importantly, assessment of their learning.

One possible method for diversifying science instruction is building and manipulating models because these are activities in which scientists are actually engaged. From a nature of science viewpoint, modeling provides a much more realistic picture of what science is than a typical problem-solving activity. Moreover, modeling is a much more robust activity because it is a creative process. To successfully build a model from a set of data requires that students understand the meaning and limitations of the data, generate an appropriate symbolic representation, and evaluate how well it works. Such an activity is almost certainly more intellectually demanding and rewarding for our students than having them work long series of plug-and-chug type problems.

Summary

This chapter describes the existence of a problem-solving mindset among many of the students enrolled in physical chemistry. Although we only studied a relatively small group of students, there are solid reasons to believe that this mindset is quite common and widespread. In the context of quantum mechanics, the existence of a problem-solving mindset has proven to be a useful tool for understanding the behaviors and difficulties that students experienced in physical chemistry classes. However, we feel that the potential impact of recognizing the existence of a problem-solving mindset is not limited to just physical chemistry, but will be applicable to a wide array of science classes.

References

1.   Thacker, B.-A.; Leff, H.; Jackson, D. *American Journal of Physics*, **2002**, *70*(3), 199.

2. Bao, L; Redish, E. (2002). *American Journal of Physics*, **2002**, *70*(3), 210-217.
3. Catalogu, E.; Robinett, R. *American Journal of Physics*, **2002**, *70*(3), 238-251.
4. Fletcher, P.; Johnston, I. *Quantum mechanics: Exploring conceptual change*. Paper presented at the annual meeting of the National Association of Research in Science Teaching, Boston, 1999.
5. Johnston, I.; Crawford, K.; Fletcher, P. *International Journal of Science Education*, **1998**, *20*, 427-446.
6. Wittmann, M.; Steinberg, R.; Redish, E. *American Journal of Physics*, **2002**, *70*(3), 218-226.
7. Bodner, G. M. Toward a unified theory of problem solving: A view from chemistry. In M. U. Smith (Ed.) *Toward a unified theory of problem solving: Views from the content domains.* (pp. 21-34) Hillesdale, NJ: Lawrence Erlbaum Associates, 1991.
8. Bodner, G. M.; Herron, J. D. Problem solving in chemistry. In J. K. Gilbert (Ed.) *Chemical education: Research-based practice.* Dordrecht: Kluwer Academic Publishers, 2002.
9. Gabel, D.; Bunce, D. Research on problem solving: Chemistry. In D. Gabel (Ed.), *Handbook of research on science teaching and learning* (pp.301-326). New York: Macmillan, 1994.
10. Bodner, G. M. *University Chemistry Education*, **2003**, *7*, 1-9.
11. Wheatley, G. MEPS Technical Report 84.01, School Mathematics and Science Center, Purdue University, West Lafayette, IN.
12. Marton, F. Phenomenography. In T. Husen & T. N. Postlethwaite (Eds.), *The international encyclopedia of education* (2nd ed., Vol. 8, pp. 4424-4429). Oxford, U. K.: Pergamon, 1994.
13. Bodner, G. M. *Journal of Chemical Education*, **1986**, *63*, 873-878.
14. Herron, J. *The chemistry classroom: Formulas for successful teaching.* Washington, D.C.: American Chemical Society., 1996.
15. Schleiermacher, E. D. *Hermeneutics: The Handwritten Manuscripts.* Missoula, MN: Scholars Press, 1997.
16. Bodner, G. M. *Journal of Chemical Education,* **2004**, *81*, 618-628.
17. Patton, M. Q. Qualitative Research and Evaluation Methods (3rd ed.). Thousand Oaks, CA: Sage Publications, 2002.
18. van Manen, M. *Researching lived experience: human science for an action sensitive pedagogy.* Albany, NY: State University of New York Press, 1990.
19. Marton, F. *Instructional Science,* **1981**, *10*, 177-200.
20. Huberman, M.; Miles, M. B. T*he qualitative researcher's companion.* Thousand Oaks, CA: Sage Publications, 2002.
21. Muhr, T. Atlas.ti (Version 4.2) [Windows]. Berlin: Scientific Software Development, 1997.

22. Bodner, G. M.; Gardner, D. E.; Briggs, M. W. Models and modeling. In N. Pienta, M. Cooper, & T. Greenbowe (Eds.) *Chemists' Guide to Effective Teaching.* Prentice-Hall: Upper Saddle River, NJ, pp. 67-76, 2005.
23. Skemp, R. *Intelligence, learning, and action.* Chichester, England: John Wiley & Sons, 1979.
24. Ausubel, D.; Novak, J.; Hanesian, H. *Educational psychology: A cognitive view* (2nd ed.). New York: Holt, Rinehart, and Winston, 1978.
25. Redish, E.; Steinberg, R. *Physics Today*, **1999**, January, 24-30.
26. Hammer, D. *Cognition and Instruction*, **1984**, *12*, 151
27. Carter, C. *The role of beliefs in general chemistry problem solving.* Unpublished doctoral dissertation, Purdue University, West Lafayette, IN, 1987.

Using Computers to Aid in the Teaching of Physical Chemistry

Chapter 10

Physical Chemistry Curriculum: Into the Future with Digital Technology

Theresa Julia Zielinski

Department of Chemistry, Medical Technology and Physics, Monmouth University, West Long Branch, NJ 07764

Traditional physical chemistry courses consist of two or three 45-hour semesters using traditional 1000+ page text books containing hundreds of exercises and problems. Typically physical chemistry courses include the study of key topics and a plethora of concepts, exploration of mathematical models of physical systems, and the use of increasingly sophisticated mathematical manipulations, all accompanied by an emphasis on scientific writing. Given the breadth of the discipline and its wide range of applications across other scientific disciplines we must ask two key questions. First, what of all available topics should an instructor include in a typical course? Second, what components of physical chemistry courses have the potential to lead to enhanced student learning and future professional growth? This paper will highlight some suggestions regarding the conundrum of too much too fast for the young physical chemistry student. Included will be examples showing project-based laboratory and lecture activities and the use of symbolic mathematics software as mechanisms for including more modern topics or more modern approaches to standard topics.

If physical chemistry is what physical chemists do, then glimpses into the contents of the *Journal of Physical Chemistry A/B* can provide some insights regarding what the curriculum might include. The A *Journal* papers are separated into five classes: 1) Dynamics and Relaxation; 2) Spectroscopy; Gaseous Clusters and Molecular Beams; 3) Kinetics, Atmospheric, and Environmental Physical Chemistry; 4) Molecular Structure, Bonding, Quantum Chemistry and General Theory; 5) and General Physical Chemistry. Part B has: 1) Surfaces and Interfaces; 2) Biophysical Chemistry; and 3) General Papers.

A review of the *Journal of Physical Chemistry A*, volume 110, issues 6 and 7, reveals that computational chemistry plays a major or supporting role in the majority of papers. Computational tools include use of large Gaussian basis sets and density functional theory, molecular mechanics, and molecular dynamics. There were quantum chemistry studies of complex reaction schemes to create detailed reaction potential energy surfaces/maps, molecular mechanics and molecular dynamics studies of larger chemical systems, and conformational analysis studies. Spectroscopic methods included photoelectron spectroscopy, microwave spectroscopy; circular dichroism, IR, UV-vis, EPR, ENDOR, and ENDOR induced EPR. The kinetics papers focused on elucidation of complex mechanisms and potential energy reaction coordinate surfaces.

Two papers reported powder pattern crystallographic results. The paper by Santos et al. (*1*) stood out from the rest because it presented a collection of more classical physical chemistry experiments. In this paper the authors described the use of micro-combustion calorimetry, Knudsen effusion to determine enthalpy of sublimation, differential scanning calorimetry, X-ray diffraction, and computed entropies. While this paper may provide some justification for including bomb calorimetry and Knudsen cell experiments in student laboratories, the use of differential scanning calorimetry and x-ray diffraction also are alternatives that would make for a crowded curriculum. Thus, how can we choose content for the first physical chemistry course that shows the currency of the discipline while maintaining the goal to teach the fundamentals and standard techniques as well?

Why a Physical Chemistry Course

While training new physical chemists is a noble ambition, most recognize that the majority of our students will not be physical chemists. This means that we are primarily training all chemistry majors in core physical chemistry concepts and skills. Physical chemistry also provides many fundamental concepts used by the other chemistry subdisciplines such as biochemistry, nanotechnology, analytical chemistry, spectroscopy, organic chemistry, inorganic chemistry, etc. Other course goals may include ensuring that our students pass the ACS Physical Chemistry exam or the graduate school physical chemistry qualifying examinations. Since these tests primarily cover the core curriculum, sticking to the standard core is a safe way to proceed. However, some instructors may want to provide a course where students become mathematically proficient at doing derivations and error analysis, while other instructors may think that students should do what the instructors did as undergraduates. These and other reasons lead us to want to consider what the real purpose of the physical chemistry course should be without disregarding curriculum standards or general academic standards. How can we move the curriculum forward to embrace more of the current literature in lecture and laboratory without sacrificing core physical chemistry concepts or losing our students along the way?

Direction Suggestion

At the ACS meeting in San Diego, Dick Zare (2) summarized the key elements to consider when changing physical chemistry course content. He first indicated that the course should start with quantum concepts from which one can determine structure and change using radiation-matter interactions. This would be followed by an integration of macroscopic observations and molecular interpretations. Since statistical thermodynamics follows comfortably after quantum mechanics, one thus has the foundation upon which to build classical thermodynamic concepts. The third component would be a treatment of chemical kinetics and reaction dynamics. Even an introductory course should stress advances in computation and the application of physical chemistry to the study of chemical processes in living things. These recommendations are broad in scope but couched in a conservative stance in that Zare recommends caution when introducing recent developments because these may hinder the learning of fundamentals that are of timeless value. Unfortunately this excellent broad outline is devoid of details and those of us who did not attend his lecture are left to imagine how to implement the key components of Zare's outline. Zare has not been the first to promote teaching quantum mechanics first. Some current texts put quantum first or come as collections with the quantum chemistry, thermodynamics, and kinetics in separate volumes that permit the instructor to choose what to put first in their courses (3-5). Implementing the Zare recommendations is not an easy task for any individual instructor. However, combinations of contributions by many authors can generate collections of materials that can meet the needs of physical chemistry teachers. Descriptions of specific examples showing the use of multi-concept context rich modules, symbolic mathematics, and computational chemistry appear in following sections of this paper.

Context Rich Instruction Modules

According to Zare, quantum chemistry should come early in the training of chemistry students. Placing quantum chemistry before thermodynamics is done by some and should be considered by all teaching faculty. During a quantum chemistry semester, one goal can be to include an introduction to modern computational chemistry using current chemistry topics and state of the art computational tools. Laboratory and lecture projects are an excellent way for accomplishing this goal. A collection of 14 projects has been developed by the Physical Chemistry Online consortium (6). This collection was initiated to respond to the need for applied physical chemistry instructional materials that has been recognized and addressed in various formats (7-10) since 1992. Four of these projects are described below.

Cl_2O_4 in the Stratosphere

The module "Cl_2O_4 in the Stratosphere" (11) uses ab initio computational chemistry with large basis sets and symmetry point group assignments to study the structure and properties of Cl_2O_4 isomers. In particular, it may be possible to predict whether Cl_2O_4 is involved in catalytic cycles that significantly deplete ozone in the stratosphere through the following reaction:

$$Cl_2O_4 + 2\,O_3 \rightarrow Cl_2O_6 + 2\,O_2 .$$

This module is an excellent example of using modern computational chemistry, classical point group assignments, isomer stability comparisons, and classical thermodynamics concepts. Energies of all species, determined with the best available ab initio Gaussian basis set, are used to compute the enthalpy of reaction for the photolysis of Cl_2O_4. The entire project is within the grasp of undergraduate students who have access to one of the basic computational chemistry software packages (Gaussian (12), Spartan (13), or Hyperchem (14)). The web site for this module contains instructor notes, hints for implementation of the project with students, a summary of computed results, and an overview of computational chemistry methods (15,16). More importantly this project can be extended or truncated by a creative colleague to fit the needs of a local curriculum. Through this activity an instructor can introduce the use of Gaussian basis sets, self-consistent field theory, molecular modeling, and thermodynamics in an integrated problem-based application. Students get a complete picture rather than a collection of unconnected concepts and processes.

Carbon Clusters

The study of carbon clusters (*17*) is another example of introducing computational chemistry and spectroscopy into the undergraduate physical chemistry curriculum. In this project students are told about a discovery made by a group of astronomers and chemists at the Kitt Peak National Observatory (*18*). Using a 4-meter telescope and Fourier transform infrared spectrometry, these scientists observed the absorption spectrum of the carbon star IRC+10216. They found a series of infrared absorption lines in the region of 2164 cm^{-1}. From the lines observed one can infer that two lines are missing between 2164.733 and 2165.870. Because this is a carbon-rich star there is the possibility that these lines are due to an infrared absorption transition in a small carbon cluster molecule - probably C_3, C_4, or C_5. The project proposes to use computational chemistry to help identify the molecule that is being observed, and then to infer some of its molecular properties from the spectra and computational results. This project is a nice accompaniment to the traditional HCl FT-IR experiment or even a substitute for it in the hands of a creative teaching scientist.

The carbon cluster project includes developing an array of physical chemistry concepts and computational chemistry techniques. After deciding on the range of isomers to consider, students use semi-empirical calculations with geometry optimization and examination of the HOMO and LUMO orbitals to choose the most likely structures to study further. High level ab initio calculations, RHF 631G(d) to UMP2/631G, with geometry optimization further refine the isomer clusters so that students can summarize bond lengths and angles of the isomers along with being able to determine their relative stabilities. Calculated ab initio vibrational frequencies along with assignment of structures to Raman or IR lines lead ultimately to the selection of the most likely cluster to account for the missing lines in the IR spectrum. The relative enthalpies and entropies obtained from the computed vibrational frequencies provide a nice link to classical thermodynamics.

For those more inclined to use environmental topics to enrich thermodynamics and kinetics parts of the physical chemistry curriculum, "Modeling Stratospheric Ozone Chemistry" and the "Contrail" projects are two examples.

Modeling Stratospheric Ozone Chemistry

In the ozone project the designers wrote:

> What sort of chemistry controls the formation of an ozone layer? What effects might man-made chemicals have on the ozone layer? These

questions are fertile territory for physical chemists. In this project we will draw upon data collected by careful experimental measurements of physical chemists working in the laboratory and in the atmosphere. Using their data, the powerful numerical methods capabilities of Mathcad, and some basic ideas from kinetics, we will explore how the concentrations of ozone and other stratospheric species change with time.

In this project students review the Chapman cycle mechanism in detail and some photochemistry concepts including the photostationary state. A key element of this project is its focus on an important chemical mechanism and the use of exploratory options for predicting ozone concentrations as a function of time while reviewing other fundamental chemical kinetics concepts. Mathcad is used as the symbolic mathematics engine for solving the requisite differential equations and ample instruction is provided to students to guide them on the use of the software in this project.

Contrails Project

In the Contrails project the developer notes the following:

Physical Chemistry is fun when natural phenomena are explained and basic chemistry principles are applied. It is especially useful when the nature of the problem involves an inter-disciplinary approach in which physical chemists play major roles in contributing key concepts and ideas. Our objective in this project is to use mathematics and physics applied to chemistry and climatology to determine the possibility of contrail formation behind the jet planes.

In this project students work collaboratively to build an understanding of the vapor pressure diagrams of ice and water and the cooling of jet exhaust. In particular students are asked to estimate the combustion chamber temperatures and pressures of a jet engine, estimate air temperature and pressure as a function of altitude, and predict cloud formation using phase diagrams, jet exhaust pressures and temperatures and weather data. Students using this module should be intrigued by the use of a simple one-component phase diagram for water along with the plot for air temperature and vapor pressure (humidity of the atmosphere) and the vapor pressure and temperature plot for jet exhaust to predict the appearance of contrails, frost in the atmosphere.

Projects in General

As the above four examples show, one can break out of the mold of standard laboratory and lecture processes by creating and using projects that span several weeks and include several fundamental concepts. Projects have been run collaboratively among participating classes from different campuses (*19-21*). This works well to support small classes by creating a critical mass of students who can contribute to the development of each other's concepts through intercollegiate collaborative learning. The projects have also been used in single classes where the number of students is large enough to support groups which serve as teams to complete different aspects of a project. Through collaborative work students can construct their understanding and more securely connect new concepts to their intellectual frameworks. This brings currency to the course, an appreciation that physical chemistry is more than just an advanced mathematics class, and may create a positive impact on long term retention. The other available projects are listed in Table I.

Using these physical chemistry projects requires some scheduling flexibility. Each project requires several weeks (4-6 weeks is common) before students reach closure. Admittedly there are few instructors willing to give a third or half a semester to a single project in an already crowded curriculum without strong reasons to do so. Before discarding the idea one should consider that none of these projects requires full class or laboratory time for any week(s) of a semester. These projects are highly modular and can be implemented concurrently with other laboratory experiments and/or lecture class work. Much of the work of a project can be done outside of class time and can substitute for some assigned homework. Thirty minutes or so at the beginning of a laboratory period or during automated data acquisition is ample for students to participate and report on the progress during a project. Closure for students can be ensured by requiring a laboratory report that can be counted as equivalent to up to two full laboratory reports out of a typical semester of seven reports so as to acknowledge the extra work involved. Multitasking permits students to complete traditional experiments and participate in longer projects. If the projects are run intercollegially then asynchronous communication is the norm and more out of class time is required to supplement lab time or class time. Intercollegiate projects require a host web site with suitable course management software for posting and checking progress across participating campuses.

A strong faculty facilitator and strong faculty support on each participating campus is required for project success. Students need constant monitoring by faculty to ensure timely completion of project assignments. Students also need instruction on how to participate and how to use their scientific voice when

Table I. Other Physical Chemistry Projects with Content Area

Project Title	Content Area
Be My Guest: Thermodynamics of Inclusion Compounds	Spectroscopy, Thermodynamics
Real Gases	Thermodynamics
Spectroscopy of I_2	Spectroscopy
Polymer Elasticity – Bungee Jumping	Materials Science, Thermodynamics
Polymers are Us	Materials Science, Thermodynamics
Laser Dyes and Molecular Modeling – The Shady Laser	Spectroscopy, Computational Chemistry
How Hot is That Flame	Thermodynamics
Computational Chemistry and Hair Dyes	Spectroscopy
Apple Enzyme Kinetics	Kinetics, Biochemistry
Chirality	Spectroscopy
The DNA Melt	BioPhysical Chemistry
Up And Down: Energy Transitions Make Lasers	Spectroscopy and Lasers

sharing ideas in writing with student colleagues on a distant campus. When implementing a project one or two older more traditional experiments should be eliminated to make room for the project and ease the burden on students. The rewards in terms of learning current physical chemistry methodology more than compensates for deletion of some older 'cherished' experiments that may be of little interest to students. In lecture the time spent on some topics can become part of a project through guided inquiry methods thus freeing up lecture time for more modern literature based topics.

"Physical Chemistry with a Purpose" created by Michelle M. Francl is another collection of project type physical chemistry materials (22). There are six modules available for student use. Each module consists of a recent paper from the literature accompanied by questions that guide the student through critically reading the paper. Follow-up questions connect the content of the paper to the topics covered in a traditional undergraduate physical chemistry course. The collection is more focused than typical case studies and unlike text books provides a clear link to the chemical literature and draws student attention to how physical chemistry is relevant to current chemical research. These modules are easily incorporated in typical classes along with homework. Strong faculty guidance is important here so that students, most of whom are not familiar with independent projects in physical chemistry, reach closure. These projects serve as a model for others interested in enriching the physical chemistry curriculum.

The use of projects is relevant when teaching physical chemistry because they expose students to the idea that physical chemistry is important to the work of all kinds of scientists: organic chemists, climatologists, kineticists, computational chemists, biochemists, etc. Physical chemists only need to be creative in designing the modules that tie the concepts together while making them intellectually stimulating and accessible to students in a guided inquiry and constructivist manner. The benefit to teaching faculty is the collegial relationships that develop during intercollegiate projects. This accompanies a more practical aspect, namely, learning how our students understand concepts as they write about them and discuss with us how to write about them. Tested and peer reviewed projects modules can be published in the *Journal of Chemical Education* as part of the Learning Communities Online Collection of teaching materials (http://www.jce.divched.org/JCEDLib/LrnCom/index.html).

Using Digital Resources in the Physical Chemistry Curriculum

Spreadsheets

Physical chemistry teachers transitioning to digital formats in physical chemistry typically start with spreadsheets especially for student laboratory reports. Analysis of large data sets and preparation of plots with spreadsheets is an easy and effective way to increase student efficiency; digitally completing the busy work of number crunching frees time for students to reflect on the concepts and focus on the meaning of the data that were collected in the lab. The use of spreadsheets also builds upon students' prior experience with this tool. The temptation here, however, is to add additional material to an already crowded curriculum. Most of the time gained by increasing the efficiency of producing reports and completing assignments should be used for developing a better understanding of topics before the addition of extra content is considered.

Spreadsheets can be used for much more than laboratory report preparation. These tools are useful as teaching and learning aids both during lecture and for homework. A recommended approach is to use well-crafted spreadsheet materials combined with guided inquiry activities that lead the student to deeper understanding of a topic. One does not need to develop such spreadsheet templates de novo because many already exist in the *JCE* DLib archives or on the WWW. Using existing spreadsheets gives any instructor a jump start toward a digitally enhanced course. Instructors can also add to the growing *JCE* collections.

Examples of well-crafted, peer-reviewed spreadsheets can be found at the *JCE* DLib WebWare site, http://www.jce.divched.org/JCEDLib/WebWare/index.html.

Several peer reviewed and open review examples include:

- Self-Consistent Field Calculations Spreadsheet (*23*)

- The Effect of Anharmonicity on Diatomic Vibration: A Spreadsheet Simulation (*24*)
- The Relation of Temperature to Energy Spreadsheet (*25*)
- Spreadsheet Methods for Point Group Theoretical Calculations (*26*)
- Interactive Spreadsheets by Coleman (*27*)
- The Photoelectric Effect by Coleman (*28*)

Finally, from the WWW there is a small collection created by Kieran Lim from Deakin University at Geelong in Victoria, Australia. This collection was created as part of the Learning and Teaching Support Network (LTSN), MathsTEAM Project (*29*). Since quantum theory is a key part of the chemical and physical sciences, it is important even for the mathematically less well prepared student such as many of those majoring in biochemistry, biological sciences, etc. to have meaningful experiences with the topic and solving the Schrödinger equation. The collection provides a spreadsheet approach, based on approximate numerical solutions and graphical descriptions of the Schrödinger equation, to develop a qualitative appreciation of quantum mechanics among students. The collection includes studies of solutions to the Schrödinger equation for various types of potential wells including the Morse potential, a triangular potential, the barrier potential and others. The active Excel and pdf files can be obtained online (*30,31*).

Symbolic Mathematics Engines

The second method to push the physical chemistry curriculum forward with technology is by using symbolic mathematics software or spreadsheets as an integral part of a course. Symbolic mathematics engines (SME) such as Mathcad, Mathematica, and Maple provide a way for faculty to give students advanced mathematical assignments, collect answers to numerical problems, and increase student skills in problem solving. To support the increasing use of SME in the physical chemistry curriculum a collection of over 100 SME documents written by over 20 teaching chemists exists at http://bluehawk.monmouth.edu/~tzielins/mathcad/index.htm. The topics in the collection are arranged in the table of contents format shown in Table II.

Approximately half of the documents in the site have been published in the *Journal of Chemical Education* where they are permanently archived, http://jchemed.chem.wisc.edu/JCEDLib/index.html. The *JCE* Digital Library Collection (*JCE* DLib) contains both peer reviewed and open access documents. Usage policy for these documents follows standard practice for *JCE*. Faculty may distribute copies of the documents to students in their classes if they have a *JCE* subscription or their campus has an IP based subscription to the *Journal*. Students have access to documents through a campus IP based subscription.

Table II. SME Content Areas in Two Columns

• Mathcad Skills	• Gases
• General Chemistry	• Classical Mechanics
• Thermodynamics	• Crystallography
• Numerical Methods	• Statistical Mechanics
• Quantum Basics	• Statistics
• Symmetry Topics	• Molecular Mechanics
• Spectroscopy	• Fourier Methods
• Kinetic Theory	• Advanced Chemistry
• Chemical Kinetics	

Why Use SME in Physical Chemistry Courses

Although physics and calculus are prerequisites for physical chemistry and the course is the first one in which numerical methods can be used to compute physical and chemical quantities from measurable data, there is a dichotomy in actual practice that extends from plug and chug type calculations to elaborate thermodynamics and quantum mechanics derivations using partial derivatives. Incorporating symbolic mathematics into the physical chemistry course provides students with more meaningful mathematical experiences, ones that lead to better understanding of the core concepts. By systematically using SMEs and SME templates throughout a course, students advance their mathematical skills and learn to appreciate mathematical models in chemistry and in particular spectroscopy. Well-crafted templates permit students to explore and discover concepts free from the drudgery of programming or numerous error-prone hand calculations and plots. There is no reason for students to use hand plots of data given the ubiquitous availability of SMEs and spreadsheets. Digital tools are best used to help students to focus on concepts and promote instructor–student discussion.

Choosing Symbolic Mathematics Applications

Since the collection contains over 100 documents faculty must choose which documents to use in their courses. The most effective mechanism is for faculty to require use of an SME from the very start of a course and build opportunities for students to learn the software as they complete homework and write laboratory reports throughout the semester. Thus one would start with a training exercise for the SME and follow with simple calculations and exercises in every homework assignment. As the semester proceeds one can introduce skills for doing derivations that support the concepts in the course. A first example could be the integration of $C_P(T)$ to get heat required to warm a sample of a material. It is important to emphasize to students that although this may seem harder at first compared to using paper and pencil, it is actually easier in the long term when they can use their own previously developed worksheets to solve new exercises or build solutions to more complex problems. A small selection of documents that would be useful in a physical chemistry curriculum follows.

Documents suitable for a quantum chemistry semester include:

- Playing with Waves
- Introductory Explorations of the Fourier Series
- Exploring Orthonormal Functions
- Blackbody Radiation
- Harmonic Oscillator Wavefunction Explorations
- Properties of the Radial Functions
- Introduction to Franck-Condon Factors (*32*)
- The Iodine Spectrum (*33*)
- Exploring the Morse Potential (*34*)
- Vibronic Spectra of Diatomic Molecules and the Birge-SponerExtrapolation (*35*)

The documents suitable for a thermodynamics course include:

- van der Waals and Redlich Kwong: Fitting Two-Parameter Equations to Gas Data
- Maxwell Distribution of Gas Molecule Velocities
- Computing a Flame Temperature
- Fitting a Polynomial to $C_P(T)$ data for Ag
- Calculating Enthalpies of Reactions
- Computing a Liquid-Vapor Phase Diagram
- Modeling Stratospheric Ozone Kinetics, Part I and II

SME use can also be required in the laboratory. All reports can be generated in digital format and calculations done with an SME, either Mathcad or Excel. The ease of interweaving text, calculations and plots in a single document makes software like Mathcad a valuable tool for preparation of student reports. Standard cut and paste from SME or spreadsheet can be used with word processing software to produce more traditional laboratory reports. A strong argument for digitally generated reports is that they can be submitted electronically as SME documents, digitally annotated and graded, and returned electronically to students. This also provides a mechanism by which faculty can archive student work for outcomes assessment or formative assessment of course materials.

Most of the documents in the *JCE* SME collection were written for Mathcad. However, the collection is expanding and adding Mathematica and Maple documents that are either original contributions or translations of existing Mathcad documents. Contributions or translations from colleagues who use Mathematica and Maple are needed to round out the collection and make it useful to a wider audience. Reviewer volunteers are also needed for documents that are submitted for publication in the *JCE* SymMath feature column.

New Directions

Physical chemistry's scope is very broad in its own right and important as a foundation for other chemistry disciplines. Our role as physical chemistry teachers is to prepare the youngest scientists in our community to be able to enter a wide variety of chemistry based careers. Given the scope of the discipline we cannot hope to cover everything that is important and/or interesting in two or three semesters. We must exercise expert critical thinking when selecting topics for both lecture and laboratory. We fail in our teaching role if we do not question the choices made by publishers and textbook authors who at present seem to be setting the agenda for course content.

Perhaps it is time for a new approach, one based more on local campus needs and faculty experience, one with greater flexibility for the instructor. This approach is now being explored through the *JCE* DLib Living Textbooks for Chemistry project, http://www.jce.divched.org/JCEDLib/LivTexts/index.html. The first entry in this area is "Quantum States of Atoms and Molecules" (*36*). This resource contains an introduction to quantum mechanics as it relates to spectroscopy, the electronic structure of atoms and molecules, and molecular properties. This resource contains active Mathcad templates and guided inquiry activities for students to use alone or in groups to increase their understanding of various concepts and develop information processing, critical thinking, and problem solving skills. With a living textbook there is opportunity for members of the teaching community to add chapters or activities. Whole sections and chapters on thermodynamics, statistical mechanics, and chemical kinetics need to be added to the living textbook. Each chapter will be peer reviewed prior to addition to the collection. The major outcome is that faculty will have available a reliable and accurate resource at minimal cost.

New directions in teaching physical chemistry should also expose students to the fundamentals of computational chemistry and the more modern methods used in the determination of rates and mechanisms of chemical reactions.

Computational chemistry is essential in a modern physical chemistry course. One approach would be to use laboratory time to have students work through a number of exercises accompanied by elaboration of the concepts in lecture or pre-laboratory discussions. Each of the major computational chemistry software packages come with workbooks or tutorials for learning the software. For example, students can learn by completing exercises in the Spartan tutorials (37). Similar approaches can be taken when using Gaussian (38) and Hyperchem (39) tutorial or exercise collections.

Chemical kinetics studies as practiced in industry are far more complex than the smattering of examples presented in typical undergraduate physical chemistry courses. The few hours we spend in typical physical chemistry courses do not include the more complex and more interesting examples in the literature because of the level of mathematics required. However, students can be introduced to more complex processes through the use of simulation software which is available via the WWW (40). Most important is that students develop an appreciation of what a mechanism is and what steps are taken to develop a mechanism for a new chemical reaction. Carefully designed simulations can help students develop these skills as reported by Houle (41).

Many other topics also can be considered for inclusion in the first-year physical chemistry course. Enzyme kinetics, the study of the solid state and crystallography (42), and polymer chemistry would be especially interesting to the biochemistry students in our classes. The choices are limited only by the creativity of faculty teaching the physical chemistry courses.

Conclusion

There are several conclusions one can draw can draw with respect to using digital technology in physical chemistry teaching. These include:

- Begin using new technology to teach physical chemistry as soon as the emergence of that technology.
- Use SME and projects from the first week of a course.
- Build student skills by constant usage of the SME throughout a course.
- Provide students with support in the form of templates and personal or online guidance.
- Use more guided inquiry and active learning to replace standard lectures in order to provide students with more meaningful learning experiences.
- Stress advances in computation throughout the chemistry curriculum.
- Embed fundamental concepts in the rich environment of projects, SMEs, and computational chemistry software.

- Search for other useful computational tools to use in the physical chemistry curriculum, develop the tool, and then share your work with the teaching community.

Transmitting a body of knowledge as one would find on a set of DVD disks or in a text book is not the goal of the physical chemistry course. Our primary goal is to empower students to be life-long learners who will continue to build upon the foundation we help them create in our classes. Our second goal is to help students build that foundation upon which they will add other topics on a need-to-know basis. The foundation should contain a sufficient breadth of the basic physical chemistry cannon to provide and enable continued learning and growth in a career. The foundation should also provide depth in several content areas to enable students to grow intellectually. Much of the fundamentals can be developed by students through highly interactive projects and SME tutorials that are grounded in concrete modern examples. After all, we learn best by doing and projects provide active guided-learning paths for students to use, therefore achieving learning by doing.

References

1. Santos, R. C.; Bernardes, C. E. S.; Diogo, H. P.; Piedade, M. F. M.; Lopes, J. N. C.; Piedade, M. E. M. d. *J. Phys. Chem A* **2006**, *110*, 2299-2307.
2. Zare, R. N., *Abstracts of Papers*. 229th National Meeting of the American Chemical American Chemical Society, San Diego, CA; American Chemical American Chemical Society: Washington, DC, 2005; CHED 722.
3. Engel, T. *Quantum Chemistry and Spectroscopy*; Pearson: San Francisco, CA, 2006.
4. McQuarrie, D. A.; Simon, J. D. *Physical Chemistry: A Molecular Approach*; University Science Books: Sausalito, CA, 1997.
5. Metiu, H. *Physical Chemistry : Quantum Mechanics*; Taylor & Francis: New York, NY, 2006.
6. Long, G. R.; Zielinski, T. J. "Physical Chemistry On-Line: Building Mastery Through Collaboration," http://bluehawk.monmouth.edu/tzielins/PCOLWEB/ChemOnLine/, (Accessed May 22, 2006).
7. Moore, R. J.; Schwenz, R. W. *J. Chem. Educ.* **1992**, *69*, 1001-1002.
8. Schwenz, R. W.; Moore, R. J. *Physical Chemistry: Developing a Dynamic Curriculum*; American Chemical Society: Washington, DC, 1993.
9. Zielinski, T. J.; Schwenz, R. W. *The Chemical Educator* **2004**, *9*, 10.1333/s00897040771a.
10. Weaver, G. C., Physical Chemistry in Practice DVD. This DVD can be

obtained from the author at Purdue University, Department of Chemistry, 560 Oval Drive, West Lafayette, IN 47907-2084.

11. Whisnant, D. M.; Lever, L. S.; Howe, J. J. *J. Chem. Educ.* **2000**, *77*, 1648.
12. Gaussian www.gaussian.com, (Accessed May 22, 2006).
13. Wavefunction www.wavefunction.com, (Accessed May, 2006).
14. Hypercube www.hyper.com, (Accessed May 22, 2006).
15. Whisnant, D. M.; Lever, L.; Howe, J. *J. Chem. Educ.* **2005**, *82*, 334.
16. Whisnant, D. M.; Lever, L.; Howe, J. J. "Cl$_2$O$_4$ in the Stratosphere," http://www.jce.divched.org/JCEDLib/LrnCom/collection/JCE2005p0334LCM/index.html, (Accessed May 23, 2006).
17. Whisnant, D. M.; Howe, J. J.; Lever, L. S. *J. Chem. Educ.* **2000**, *77*, 199-201.
18. Bernath, P. F.; Hinkle, K. H.; Keady, J. J. *Science* **1989**, *244*, 562.
19. Slocum, L. E.; Towns, M. H.; Zielinski, T. J. *J. Chem. Educ.* **2004**, *81*, 1058-1065.
20. Towns, M. H.; Kreke, K.; Sauder, D.; Stout, R.; Long, G.; Zielinski, T. J. *J. Chem. Educ.* **1998**, *75*, 1653-1657.
21. Towns, M.; Sauder, D.; Whisnant, D.; Zielinski, T. J. *J. Chem. Educ.* **2001**, *78*, 414-415.
22. Francl, M. M. "Physical Chemistry with a Purpose," http://www.brynmawr.edu/Acads/Chem/NSFpchem/, (Accessed May 22, 2006).
23. Hoffman, G. G. *J. Chem. Educ.* **2005**, *82*, 1418.
24. Kieran F. Lim; Coleman, W. F. *J. Chem. Educ.* **2005**, *82*, 1263.
25. King, C. *J. Chem. Educ* **2005**, *82*, 861.
26. Vitz, E. *J. Chem. Educ* **2002**, *79*, 896.
27. Coleman, W. F. "Interactive Spreadsheets," http://www.jce.divched.org/JCEDLib/WebWare/collection/open/JCEWWOR025/index.html, (Accessed March, 2006).
28. Coleman, W. F. "Photoelectric Effect," http://www.jce.divched.org/JCEDLib/WebWare/collection/open/JCEWWOR006/index.html, (Accessed May 22, 2006).
29. Hirst, C., Ed. *Maths for Engineering and Science*; LTSN MathsTEAM Project, LTSN Maths, Stats & OR Network: Edgbaston (UK), 2003.
30. Lim, K. F. "Using spreadsheets to teach quantum theory to students with weak calculus backgrounds," http://www.mathcentre.ac.uk/resources/casestudies/mathsteam/lim.pdf, (Accessed May 22, 2006).
31. Lim, K. F. In *Maths for Engineering and Science*; Hirst, C., Ed.; LTSN MathsTEAM Project, LTSN Maths, Stats & OR Network: Edgbaston (UK), 2003, pp 24-25.
32. Zielinski, T. J.; Shalhoub, G. M. *J. Chem. Educ.* **1998**, *75*, 1192.
33. Long, G.; Zielinski, T. J. *J. Chem. Educ.* **1998**, *75*, 1192-1192.
34. Zielinski, T. J. *J. Chem. Educ.* **1998**, *75*, 1191.

35. Standard, J. M.; Clark, B. K. *J. Chem. Educ.* **1999**, *76*, 1363-1366.
36. Hanson, D. M.; Harvey, E.; Sweeney, R.; Zielinski, T. J. "Quantum States of Atoms and Molecules," http://www.jce.divched.org/JCEDLib/LivTexts/ pChem/JCE2005p1880_2LTXT/index.html, (Accessed May 22, 2006).
37. Wavefunction *Spartan '04 Tutorial and User's Guide*; Wavefunction, Inc.: Irvine, CA, 2003.
38. Foresman, J. B.; Frisch, A. *Exploring Chemistry with Electronic Structure Methods*; 2nd ed.; Gaussian Inc.301: Pittsburgh PA, 1996.
39. Caffery, M. L.; Dobosh, P. A.; Richardson, D. M. *Laboratory Exercises Using HyperChem*; Hypercube, Inc.: Gainesville FL, 1998.
40. Hinsberg, W.; Houle, F. "Chemical Kinetics Simulation," http://www.almaden.ibm.com/st/computational_science/ck/msim/?overview, (Accessed May 22, 2006).
41. Houle, F. A.; Hinsberg, W. D. In *Annual Reports in Computational Chemistry*; Spellmeyer, D. C., Ed.; Elsevier: New York, 2006; Vol. 2.
42. Rhodes, G. *Crystallography Made Crystal Clear: A Guide for Users of Macromolecular Models*; 3rd ed.; Academic Press: New York, 2006.

Chapter 11

"Partial Derivatives: Are You Kidding?": Teaching Thermodynamics Using *Virtual Substance*

Chrystal D. Bruce[1], Carribeth L. Bliem[2], and John M. Papanikolas[2]

[1]Erskine College, 2 Washington Street, Due West, SC 29639
[2]Department of Chemistry, CB 3290, Venable and Kenan Laboratories, University of North Carolina, Chapel Hill, NC 27599

Inquiry-driven student learning of thermodynamics is achieved with the implementation of the *Virtual Substance* molecular dynamics simulation program. Problem solving and critical thinking skills are developed by students through the completion of the modules described in this chapter. In addition, *Virtual Substance* is shown to be exceptionally numerically accurate as a tool for doing authentic science.

Introduction

As instructors of physical chemistry, we all too often hear comments like the one in the title of this chapter. The prevailing attitude with which many students approach physical chemistry is one of fear for their academic survival. The concerns they have are not so much with the chemistry but with the math. For some students, the physical chemistry course is the first time they must authentically apply advanced mathematical concepts to solve real problems. Instructors can facilitate this transformation with exercises that demand critical thinking skills, but developing such materials requires extensive time and talent. The *Virtual Substance* molecular dynamics program is a powerful tool in our efforts as educators to help students successfully transition from abstract knowledge of mathematics to its application to scientific concepts.

Currently, many software programs are available that provide opportunities to visualize some of the more complicated concepts in the undergraduate chemistry curriculum. For instance, commercially available quantum mechanical software programs such as Gaussian (*1*), Spartan (*2*), Hyperchem (*3*), or CAChe (*4*) allow students to build molecules, calculate their vibrational frequencies, and observe the vibrational modes as the molecules bend or stretch. Mathematical programs like Mathcad (*5*) or Mathematica (*6*) enable students to integrate complex equations, allowing them to see the myriad calculations that comprise molecular orbital theory or determine values of the Virial coefficients. Indeed, the incorporation of these and similar programs has changed the way college physical chemistry is taught and the types of problems undergraduates can tackle.

Sometimes, though, these programs can seem like a "black box" to students; while they can produce pretty pictures or the values of constants, students do not gain the conceptual understanding to analyze those pictures and values. The proper use of these programs requires insight and much thought into the design of assignments that guide inquiry and analysis so as not to remove the thinking that the students must do. Further, few of these programs directly address thermodynamics, a subject that begs for interesting material, especially from the students' point of view.

Virtual Substance, by contrast, offers a host of possibilities that can bring thermodynamic concepts to life. It offers the user an interface with a learning curve that quickly removes the element of the "black box", yet the resulting analysis of the generated data requires serious thought and understanding of the concepts to make sense. After some experience with the program and progression through the course material, the connection between molecular-level forces and thermodynamic properties becomes intuitive to students. *Virtual Substance* transforms physical chemistry concepts such as radial distribution functions, phase transitions, and real gas (versus ideal) behavior from abstract mathematics to real-world understanding, and it does so with exceptional numerical accuracy.

A framework for using computer models

As physical chemists, our goal is to describe and predict chemical properties of real substances from a molecular perspective. How are pressure and volume related for argon gas, for instance? The straightforward way to *describe* argon is to perform experiments in which the pressure is varied and record the volumes that result, leading to the inverse relationship originally observed by Boyle.

The ability to *predict* properties of a substance requires a thorough understanding of what is occurring at the microscopic level. Using this knowledge, one can construct an analytical expression that relates the macroscopic properties of the gas. This expression, an equation of state (EOS), describes the thermodynamic quantities of the substance. The ideal gas law is one such expression, and the van der Waals equation is another. Both expressions describe a the pressure of a gas relative to its volume, but they incorporate different definitions of force between atoms to do so. Not surprisingly, these two equations of state predict different pressures.

By comparing experimental data with equations of state, we are able to test the analytical expression for accuracy. In this way, we find, for example, that the ideal gas law works well when the system exhibits low pressure or large volume but falters at high densities. We may choose to refine the equation to more accurately predict experimental outcomes or we may see the differences as a trade-off for the simplicity of the EOS.

Another avenue for *predicting* properties of a substance can be obtained by creating a computer model of the substance, a *virtual substance*, and observing its properties. Key to this modeling is our choice of the interactions between atoms. In the model, we can decide what type of interaction we want to include between the atoms. The simulation shows us exactly what is going on between particles *for the given intermolecular potential*. It provides a window to the microscopic level that not only informs our understanding of the gas but actually adds knowledge regarding the physics of the system – knowledge not available from physical experiments.

The comparison of computer models with experimental data, then, tests the accuracy of the model. Assuming good agreement, we can take our analysis one step further: by comparing equations of state with computer simulations, we test the assumptions implicit in the theories that lead to the EOS. That is, we shed light on what parameters in the analytical expression give rise to observations in the computer simulation. We can assess which underlying assumptions in the EOS constrain its usability.

What is Virtual Substance?

In essence, *Virtual Substance* can be used as a new way of incorporating software in chemistry - a tool for doing 'real science' on a desktop computer. *Virtual Substance* is a molecular dynamics simulation program that generates the properties of a substance in the solid, liquid, and gaseous phases by describing the substance as a collection of individual particles. The easy-to-learn software has as input an intermolecular (interatomic) potential that is used to calculate forces on individual atoms. It then integrates Newton's equations of motion over time in order to observe the dynamics of the system (i.e., the positions and velocities of all the particles as a function of time). The program provides a three-dimensional view of the system that shows the motion of the particles during the course of the calculation. The link to thermodynamics is made by then using the forces, positions, and velocities to calculate the pressure (P), temperature (T), and total energy (E). When P, T, and E are combined with the volume (V), one obtains the thermodynamic state of the system.

Figure 1 shows the initial user interface and an example of a generated virtual substance. Depending on the parameters selected, *Virtual Substance* can model gases (ideal and real), liquids, solids, and polymers. Available intermolecular potentials include the soft sphere model (repulsive forces), the Lennard-Jones model (attractive and repulsive forces), and the Lennard-Jones-with-FENE (finite extension nonlinear elastic) model for examining the statistical mechanics of polymeric chains. (7) The program is parameterized for the noble gases He, Ne, Ar, Kr, and Xe, and it has the capability to examine user-defined substances. Users set the number of particles in the system, which is limited by the system memory and processing power of the user's computer. Calculations involving 100-200 atoms are tractable on most desktop computers.

Virtual Substance simulations can be run using either periodic boundary conditions or fixed walls. In a fixed wall calculation, the simulation box has physical walls. The resulting system is a "droplet", albeit a rectangular droplet, and as a result, the thermodynamic properties differ from those of the bulk substance. One can create a bulk substance by using periodic boundary conditions where the simulation box interacts with copies of itself repeated in three dimensions. (8) This approximates the extended nature of a bulk material, making it appear infinite, and results in better accuracy in the calculation of thermodynamic properties.

Once users have generated a substance, they then choose a simulation type: constant energy (E), constant volume and temperature (V,T), or constant pressure and temperature (P,T). Additional simulation conditions that must be set are the timestep for the integration of the equations of motion (in femtoseconds) and the number of steps. Figure 2 shows the simulation interface for choosing these parameters. Throughout the simulation, instantaneous values of T, P, V and E are shown. At the completion of the simulation, average values of the temperature, pressure, volume, total energy, kinetic energy, and potential energy are reported. Students can also generate a list of x,y, and z coordinates or velocities for each atom using the configuration tab.

Another useful feature is the command script. This built-in command language enables users to automatically perform calculations at a series of P, V, or T values; that is, to submit a batch of calculations and save every set of results. Thus users can parameterize and initiate a set of calculations and walk away; when they return, the data is ready to be analyzed – the real challenge!

Thermodynamic concepts studied using *Virtual Substance*

Virtual Substance can be used as supplemental homework assignments in the physical chemistry lecture course or as laboratory modules for a physical

a)

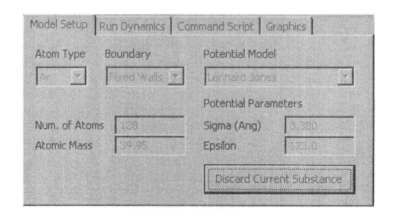

b)

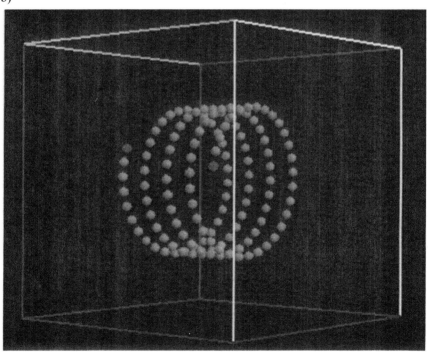

*Figure 1. a) Virtual Substance model building interface
b) Sample generated substance.*

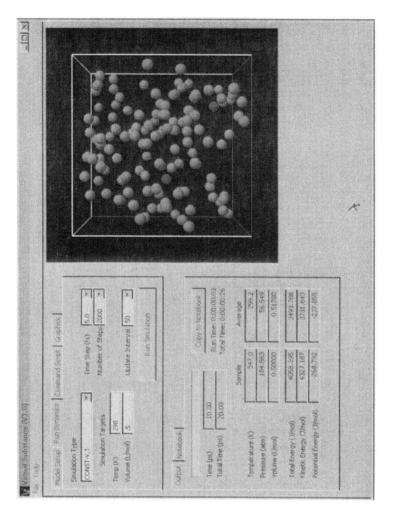

Figure 2. Virtual Substance simulation interface

chemistry lab course. At Erskine College, *Virtual Substance* has been used solely in the laboratory portion of the course. Students are required to write formal reports based on the Virtual experiments. By requiring significant quantities of data analysis and extensive write-ups, the students internalize what they learn using *Virtual Substance* possibly more than any other projects/assignments/material they complete. The assignments are designed to build upon each other: the first, an exploration of the ideal gas law to determine the value of the gas constant, R, gives the students an opportunity to learn the *Virtual Substance* program without worrying about the methods involved in the molecular dynamics simulation. By the end of the semester, the students have had enough experience with *Virtual Substance*, molecular dynamics, and thermodynamics to complete an independent project. (The partial list of possible topics given to the students is shown in Table 1.) Each student submits a proposal, completes a set of simulations, presents his or her results to the department, and writes a formal report. Although having all of these projects going on at once at the end of the semester can be hectic for the professor, this process is an invaluable learning experience for the students.

Ideal Gas Law

The purpose of this first assignment is to orient the students to the *Virtual Substance* program in the context of familiar theory. A detailed procedure leads students through a series of steps including simulation conditions but not data analysis. Students are asked to plot the data in some form to acquire a value for the gas constant, R. Actual student findings yield average values of 0.0825 L atm mol^{-1} K^{-1} for Argon and Xenon under either constant pressure or constant volume conditions. An important result of this work is the agreement between the measured value of the gas constant and its experimentally accepted value. For instructors, the use of *Virtual Substance* as an inquiry-driven, problem-solving tool is appealing. For students, numerical accuracy is extremely important. It gives them a sense of accomplishment and a guideline for determining where a mistake has been made if they do not get "the right answer". Now they can test the question "Is my answer reasonable?" because they know the result they should achieve.

Real Gas Behavior

One of the over-arching goals of physical chemistry is to explain real systems by building upon what we know about ideal systems and examining the limitations of those idealized models. The study of real gas behavior using *Virtual Substance* is one of the most eye-opening assignments for the students.

In this lab, the students determine the compression factor, (9) $Z = PV/nRT$, for Argon using the hard sphere model, the soft sphere model, and the Lennard-Jones model and compare those results to the compression factor calculated using the van der Waals equation of state and experimental data obtained from the NIST (10) web site. Figure 3 shows representative results from these experiments. The numerical accuracy of the *Virtual Substance* program is reflected by the mapping of the Lennard-Jones simulation data exactly onto the NIST data as seen in Figure 3.

One particularly powerful insight students gain from this assignment is the limitations of the van der Waals equation of state. Often in undergraduate chemistry courses, the van der Waals equation is presented as the universal correction to the ideal gas law, perhaps owing to its straightforwardness and the ease with which it can be understood. Recognizing its limitations leads students to consider other equations of state, where each expression has its own set of assumptions. While students are initially uneasy with the notion that the van der Waals equation has drawbacks and that decisions about which EOS to use depends on the system or context, this unease is not uncommon in the execution of real science.

Thermodynamic Properties

The final guided assignment that students at Erskine College perform is to calculate the expansion coefficient, α, and isothermal compressibility, k_T, for Argon behaving as an ideal gas. The word *guided* is used rather loosely in this case. The students are instructed to combine their knowledge of α and k_T with their experience using *Virtual Substance* to determine these values and compare them with theoretically predicted values. After perusing the textbook (9) and reminding themselves that

$$k_T = -(1/V)(\partial V/\partial P)_{T,n} \text{ and } \alpha = (1/V)(\partial V/\partial T)_{P,n},$$

students must recognize what all the symbols in these equations mean and then relate them to the simulation parameters in *Virtual Substance*. To instructors, this may not seem like a barrier, but in truth, students must integrate two concepts: the "constant" in partial derivatives and the "constant" in a simulation. In addition, they must determine what data they require in order to achieve the desired result and how to collect it in *Virtual Substance*. Given the program's straightforward interface, an unthinking student can quickly amass a large volume of data. However, when (and not if) this happens, she becomes overwhelmed by the sheer volume of output, much of which may be completely unnecessary in the context of the particular experiment. Proper execution of this

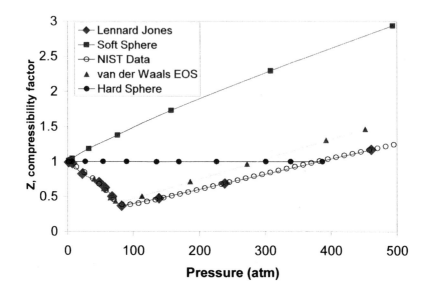

Figure 3. Student results from real gas experiments.

Table 1. Topics that can be investigated with *Virtual Substance*

Kinetic Theory of Gases	Velocity Distribution Speed Distribution Collision Frequency Mean Free Path	Determine *average* and RMS speeds from data Plot radial distribution functions
Real Gases	P(V)	Identify limit(s) of ideal behavior Determine critical constants
Energy and Enthalpy	E(T) H(T) E(V) H(P)	Determine C_V and C_P from data Determine ΔH_{vap} and ΔH_{fus} Internal Pressure Combine with C_P to get Joule-Thomson coefficient
Entropy	C_P(T)	Calculate S_m(T) Calculate ΔS_{fus} and ΔS_{vap}
Internal Pressure	P(T)	Confirm that $\pi_T = T(\partial P/\partial T)_V - P$
Phase Diagrams	Construct phase diagram from T_{fus} and T_{vap} measurements	Phase transitions Identification of phase boundaries Comparison with Clausius-Clapeyron equation
Polymer Modeling	Output x,y,z coordinates	Track end-to-end distances

assignment forces students to think critically before, during and after the experiment, that is, to develop problem solving skills - the ultimate goal of educators.

Once students have discovered what simulations to conduct and what data to collect, they must next assess how to extract the derivative. For the determination of α, a plot of volume versus temperature produces a line; most students make the connection between the slope of a line and the derivative. The interesting part occurs when they plot volume versus pressure and see a curve. Now what? While it may take a bit of time, most students realize eventually that they can fit a power series to the curve and take the derivative of the resulting function. This assignment reinforces the concepts students have been studying in the lecture portion of the course and provides an opportunity to see the applicability of the mathematics previously regarded as off-putting or of theoretical use alone.

Problem Solving Skills

Authentic learning occurs when a student uses previous knowledge to explain new information, that is, when existing knowledge and new experiences are synthesized by the student. Unfortunately, this type of learning rarely takes place in the passive setting of a lecture. It *can* occur while students are doing homework, and it *sometimes* occurs in lab, but it is often accompanied by much frustration on the part of the student. Too often, commercially available textbooks, problems, and lab protocols do not provoke thought necessary to achieve this type of learning. *Virtual Substance* modules, in contrast, are designed to foster learning physical chemistry and to develop problem solving skills by doing authentic science. By providing a protocol that offers a set of research questions rather than a step-by-step approach to working through experiments, students must consider the goals of the assignment and then devise experiments to satisfy those goals. Students generally are uncomfortable with this approach at first; they prefer that instructors and protocols provide the thinking for them. However, this facet of the *Virtual Substance* modules is exactly what encourages authentic learning.

Yet this approach does not reduce the responsibility of the instructor. In fact, the difficulty students face in finding "the answer" is almost matched by the discipline the instructor must show in not giving it to them. As mentioned in the descriptions of the *Virtual Substance* assignments, students will have to plan their attack and work through problems they encounter. This is the scientific process in practice. The task of the instructor or TA then becomes to guide the student through the process to a satisfying end result. Rather than being the keeper and giver of the knowledge, instructors (and TAs) must encourage and direct students so that they discover the knowledge themselves. It's not as easy as it sounds!

Student Comments and Evaluations

The best evaluation of student learning is application. With *Virtual Substance*, students are required to integrate and apply concepts they have acquired in various courses at different times in their education to solve problems. After completing any of the *Virtual Substance* assignments, the students often understand a concept so well that they continue to apply the knowledge they acquired in their understand of new material we encounter. For example, after working with the hard sphere model, our discussion of collisions and the kinetic theory of gases could be much deeper because the students had an intuitive understanding of the model.

Some feedback given by students include these comments:

*   It helped without my realizing it was helping. To hear the words "partial derivative" sends chills down my spine but I remember the math for *Virtual Substance* being easier.
*   I'm a big fan of figuring things out for myself. Because *Virtual Substance* allowed me to do my own thing I feel like I understood better what I was doing because it was me doing all the work from the beginning.
*   It gave me a better understanding of what was really going on in the "real" experiments in lab. *Virtual Substance* allowed you to "see inside" the container, visualizing the "actual gas particles."
*   It really helped to develop visual images of things that we don't ordinarily see (i.e. Gas expansions and such)

In addition, the average score for Erskine College students on the Thermodynamics portion of the American Chemical Society standardized exam has increased by 26.4% to 22.5 out of 40 (22 students) since the introduction of *Virtual Substance* from 17.8 out of 40 (24 students) during the years prior to the use of *Virtual Substance*. Students are demonstrating mastery of questions directly related to concepts examined using *Virtual Substance* assignments. While these numbers are based on a small sample size, they are encouraging for instructors who appreciate the value of using hands-on learning opportunities.

Conclusions

The *Virtual Substance* program can reinvigorate the learning of a broad set of thermodynamic topics as seen in Table 1, the list given to students for potential independent projects. It is an inquiry-driven tool that reinforces concepts while promoting critical thinking and developing problem solving skills with results that are extremely accurate. Other similar software programs that are currently available, while packed with many bells and whistles, are not as accurate, as user-friendly or as thought-provoking to use. The scientific process required to make sense of virtual experiments prepares students for future scientific endeavors whether they are headed to graduate school, industry, or other positions. Through the completion of *Virtual Substance* modules, students gain an understanding of a research question, learn to apply their prior "book" knowledge to design experiments, and then critically analyze their results to answer the question, all invaluable tools in the scientific arsenal. For teachers of physical chemistry, *Virtual Substance* is an extremely powerful tool to add to their collection of materials that emphasize physical chemistry concepts in general and thermodynamics specifically. Although incorporating *Virtual Substance* into the curriculum requires patience and a willingness to let students "cogitate" on the part of the instructor, the rewards are in students' advanced understanding of how mathematical concepts directly connect to chemical properties. In fact, some students even appreciate the beauty of those once-dreaded partial derivatives.

Currently, the Virtual Substance software program is available for free download at http://www.unc.edu/~jpapanik/ VirtualSubstance/VGMain.htm. Along with a concise manual, the website offers modules that address the thermodynamic concepts of the ideal gas constant, real gas behavior, and calculation of heat capacities. At this time, additional modules are available by contacting the authors; topics include radial distribution function of condensed phases, structure of polymer chains, phase transitions, and thermodynamic properties (isothermal compressibility and expansion coefficients). Because the Virtual Substance program is a work in progress, modules are being added to the website on an ongoing basis. We encourage interested users to check the website regularly and register to be notified of additions and upgrades to the program. Please feel free to contact us with questions, comments, and suggestions.

Acknowledgements

CDB would like to thank the Bell Grant fund at Erskine College for summer support. The development of the *Virtual Substance* software program was supported in part by NSF grant CHE-0301266 awarded to JMP. We also want to thank our students for participating in this endeavor.

References

1. Gaussian 03 is a registered product of Gaussian, Inc., 340 Quinnipiac St Bldg 40, Wallingford, CT 06492.
2. Spartan is a registered product of Wavefunction, Inc. 18401 Von Karman Avenue, Suite 370 Irvine, CA 92612
3. Hyperchem is a registered product of Hypercube, Inc., 1115 NW 4th Street, Gainesville, FL 32601
4. CAChe is a registered product of Fujitsu America, Inc., 15244 NW Greenbrier Parkway Beaverton, OR 97006-5764
5. MathCad is a registered product of MathSoft, Inc., 101 Main St., Cambridge, MA 02142.
6. Mathematica is a registered product of Wolfram Research Inc., 100 Trade Center Drive, Champaign, IL 61820.
7. Koplik, J.; Pal, S.; Banavar, J. R. *Phys. Rev. E: Stat., Nonlin., Soft Matter Phys.* **2002**, *65*, 021504/1-021504/14.
8. Allen, M.P.; Tildesley, D.J. Computer Simulation of liquids; Clarendon Press: Oxford, 1987.
9. Atkins, P.W.; dePaula, J. *Physical Chemistry*, 7th edition;W.H. Freeman and Company: New York, 2002.
10. Thermophysical Properties of Fluid Systems, National Institute of Standards and Technology http://webbook.nist.gov/ chemistry/fluid/ (accessed February 7, 2006).

Chapter 12

Molecular-Level Simulations as a Chemistry Teaching Tool

Jurgen Schnitker

Wavefunction, Inc., 18401 Von Karman #370, Irvine, CA 92612

Simulations of chemical systems can provide students with authentic laboratory experiences within the confines of their regular coursework. In the most common type of simulation, the focus is on the manipulation of virtual equipment or toy models of chemical processes. In a second type, here labeled "atomistic simulations," the main challenge for students is the analysis of the visual and numerical data that are generated by true models of nature. The power of the atomistic approach lends itself perfectly to the quantitative character of physical chemistry. While computationally demanding, the corresponding simulations no longer require resources other than those available on even the most modest of today's machines. With pedagogical as well as motivational value for students, atomistic simulations are poised to become a routine teaching tool.

Curriculum developers as well as textbook publishers have in recent years increasingly taken to the use of simulations for the teaching of basic science concepts. Not too long ago mostly delivered on CDs, the corresponding software is now often browser-based and available online. In fact, simulations can be a very effective tool to provide students with learning experiences that are both powerful and lasting. As outlined in the following, however, the term "simulation" is used in connection with some widely differing approaches. While all can be of value in the classroom, the associated learning objectives are quite different.

Science Media

The dictionary defines simulation as "the representation of the behavior or characteristics of one system through the use of another system" (*1*). The use of simulations in chemistry is prevalent enough to have been the subject of educational research (*2-5*). In a lot of the corresponding software, the objects being represented are the chemicals and pieces of equipment of an actual laboratory or, equally common, cartoon-style abstractions of chemical concepts. Students thus get to manipulate simulated glassware, prepare solutions by dragging icons, or shuffle electrons to satisfy the rules of quantum mechanics. In software for beginners, they may on occasion even cause fake explosions.

The simulations are developed through storyboarding and typically follow a very tight script. In technically advanced renditions (using Flash or Java applets), the students may engage in nontrivial and intellectually stimulating interactions. In a more traditional format, the delivery is through media players (such as QuickTime), leaving the student mostly as a passive observer. The term "animated narrative vignette" has been used to characterize this type of learning experience (*6*). Whether involving true student interaction or not, we will use the term "science media" for simulations and animations that are focused in the indicated way on one topic at a time.

The physical chemistry topics treated by science media span a wide range, from the photoelectric effect to the hydrogen spectrum, and from the Carnot cycle to the partition function. This is in fact the major appeal of the whole approach: the pedagogically narrow learning objectives of teaching standards and standardized examinations can be catered to with custom software that always highlights precisely the item being assessed.

In conclusion, the strength of science media is that they can potentially provide very succinct, albeit cartoonist treatments of a large number of topics. Their weakness is that they will do, and *only* do, whatever the designer or programmer had in mind. Essentially, science media are a 21st century, computer-based version of the textbook illustration.

Science-Based Learning Environments

A fundamentally different type of simulation is offered by science-based learning environments. Such environments incorporate some general-purpose mathematical engine that either represents nature directly or that can be programmed to represent nature. Examples are "Mathematica" (7) and some similar programs (8-9) for general analytical modeling in the physical sciences and "Interactive Physics" for introductory classical mechanics (10). Mathematica and Interactive Physics can be applied to countless topics, as opposed to the narrow focus of Flash-based simulations. Even more importantly, Mathematica and Interactive Physics are open-ended in that the software may accommodate unscripted inquiries and follow-up questions.

An equivalent example in the field of chemistry is molecular modeling that is based on the techniques of quantum chemistry. Several widely distributed programs are available (11-17), with "Spartan" (12) being the dominant program in organic chemistry teaching laboratories for undergraduates. Like all molecular modeling programs, Spartan is by design open-ended. Similar to Mathematica, the program is used in both research and education

Another chemistry-specific learning environment is the dedicated teaching program "Odyssey" (18). Making use of an atomistic simulation engine similar to that found in other programs (19-32), Odyssey is primarily aimed at introductory and general chemistry courses. However, the software is also applicable to many areas of physical chemistry as basic thermodynamic properties as well as more advanced properties such as collision densities and free energies can be calculated and analyzed.

Mathematica, Interactive Physics, Spartan, and Odyssey are all available with protocols and content (sometimes from third parties) for subject-specific topics. The learning objectives, however, go beyond those of specific content. Being very different from memorization aids, method-based learning environments have at least the *potential* to familiarize students with the process and the principles of science. It is precisely the use of a multi-purpose tool that makes students invoke the generic reasoning skills, both qualitatively and quantitatively, that are the hallmark of a scientifically literate mind. Sometimes there is value in using fewer applications for more purposes!

To summarize, science-based learning environments use storyboarding of topics not as a means in itself; instead, they address the subject matter via the predictive power of an underlying general engine. As tools that generate "real" data and with the ability to handle unscripted queries, science-based software environments provide a 21st century version of the laboratory experience.

Atomistic Simulations

Making connections between macroscopic and molecular phenomena is the essence of learning chemistry. Atomistic simulations probably embody this connection better than any other method. Matter is represented at a level where individual atoms are the smallest units, not electrons as in the case of quantum chemical modeling. Given the level of description, hundreds or even thousands of atoms can be handled with off-the-shelf computers and—most importantly—student-owned laptops. As a consequence, simulations of "real" molecules (such as 1-octanol or polyalanine) and "real" samples of bulk matter (such as liquid water or syngas) are feasible on a routine basis.

Exploiting the principles of statistical mechanics, atomistic simulations allow for the calculation of macroscopically measurable properties from microscopic interactions. Structural quantities (such as intra- and intermolecular distances) as well as thermodynamic quantities (such as heat capacities) can be obtained. If the statistical sampling is carried out using the technique of molecular dynamics, then dynamic quantities (such as transport coefficients) can be calculated. Since electronic properties are beyond the scope of the method, the atomistic simulation approach is primarily applicable to the "thermodynamics" half of the standard physical chemistry curriculum.

The recently introduced program Odyssey (18) makes atomistic simulations transparently available to students and instructors. At the core of the program is a molecular dynamics simulation engine that is versatile enough to handle a great variety of systems from almost all fields of chemistry. Chemistry content is directly embedded in the user interface. The presentation style is such that any given simulation is invoked as a teaching aid, rather than as a goal in itself. For this approach to work, it is essential that the user is completely shielded from technical setup issues. In fact, many other molecular dynamics programs are available (19-32), but there is none whose scope is as wide as that of Odyssey while avoiding all method-related jargon.

All Odyssey simulations are carried out interactively, rather than through the batch-style job submission common in research software. Simulations can be initiated with samples from a large repository of pre-built systems or with samples that are built from scratch using an integrated model kit. In either case, the computer draws a fully three-dimensional, manipulable, and customizable picture. After starting the simulation (typically at room temperature), the picture gets periodically updated as molecular change occurs. The result is that students can literally see chemistry happen in front of their eyes. Like a giant microscope, Odyssey provides students with a window into the molecular world.

The physical description of the simulated systems in Odyssey is via classical potential functions that have been developed for research applications. In many areas relevant to teaching, the description is at least qualitatively correct. This is all that is required from a pedagogical standpoint. Nevertheless, the models do fail on occasion, even qualitatively. Rather than being a drawback, this can well be considered a compelling illustration of the fact that eventually *all* scientific models have intrinsic limitations. As teachers of science (rather than of scientific facts), we should be conveying this to our students in any case: Going back to the laboratory is eventually the only way to find out!

Among the pedagogically most appealing features of Odyssey is the ability to influence simulations as they occur. Parameters such as the temperature, the available volume, and the system composition can be changed at any time—not because it was "all set-up" for a particular topic, but because of the fundamental power of the underlying method. Inquiry-based learning becomes possible on a much broader scale than in science media software.

Odyssey makes use not only of live calculations of bulk matter samples, but also includes (in an archived form) quantum chemical models such as calculated atomic orbitals. As a consequence, coverage of select topics from many areas of the introductory chemistry curriculum is achieved. While the program has been adopted by several hundred institutions to date, systematic field studies regarding its pedagogical effectiveness have not yet been published. Anecdotal evidence indicates that the package is appreciated for its uniqueness as well as for the significant "cool-factor" with which it is perceived by students. Many instructors have also commented that they gained new insights themselves by working with the software inside as well as outside the classroom.

Physical Chemistry Applications

The following examples involve Odyssey simulations of a few minutes duration each that are carried out under active control of the student with full three-dimensional visualization of all molecular motion.

As an example of the usefulness of molecular visualization, Figure 1 shows a 128-molecule sample of ordinary hexagonal ice. The molecules are drawn in the "Space Filling" model style, commonly known to provide a reasonable representation of the effective size of most molecules. Figure 1a shows the ideal lattice structure (T = O K). It can be seen that the structure is exceptionally open, with "channels" that permeate the entire lattice. Essentially, the picture provides a hands-on molecular illustration of the uniqueness of water (the density of the solid is so low that it actually floats on the liquid).

Does thermal motion make a difference for this aspect of the structure of ice? Figure 1b shows a snapshot from a simulation at finite temperature, prior to melting. While the perfect molecular alignments of the ideal lattice have been lost, the picture still shows discernible "channels": molecules in solids do move, but this motion does not affect the overall symmetry.

The next few examples relate to the kinetic theory approach of physical chemistry. Figures 2 and 3 show the kinetic energy distribution for a room temperature sample of 80 carbon monoxide molecules (P ~10 atm). The obtained data lend themselves to making a few important points about the interpretation of histograms. Histograms are just a special type of plot, and Odyssey can be set up to calculate and display simultaneously as many plots as

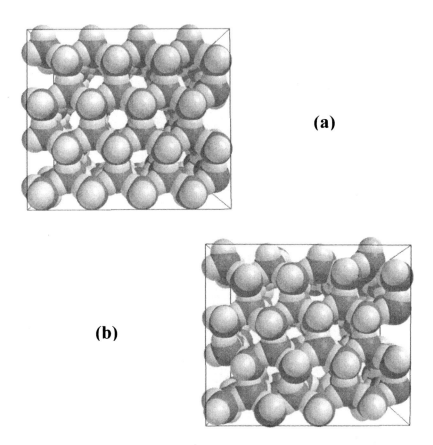

(a)

(b)

Figure 1. Molecular structure of hexagonal ice. Ideal lattice structure (a) vs. structure in the presence of thermal excitations prior to melting (b).

desired. In this case, the same histogram has been calculated three times, but with different bin widths. For a "reasonable" bin width (Figure 2), we find a behavior that is compatible with the prediction from basic kinetic theory and a corresponding illustration in a textbook:

$$\text{probability} \sim E^{1/2} e^{-E/kT}$$

For an excessively small bin width (Figure 3a), we still confirm the expected behavior, but now the distribution also shows some spurious peaks. For a very large bin width (Figure 3b), a more dramatic change can be seen: the distribution now looks qualitatively different, as the initial peak has been completely lost—the simulated data seem to contradict the expectation from analytical theory!

The observed behavior is of course easily rationalized, but students still need to *learn* that scientific data cannot be analyzed without critical judgment. While this message is conveyed by any good laboratory, it is ignored by science media-type software that is focused on making the student memorize equations and facts.

Continuing with kinetic-molecular theory, Figure 4 shows two simultaneously calculated histograms for a sample of syngas that contains 50

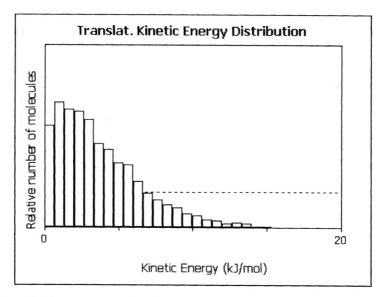

Figure 2. Kinetic energy distribution of an 80 molecule sample of CO (g) at 25 °C and ~10 atm, calculated with the teaching program Odyssey.

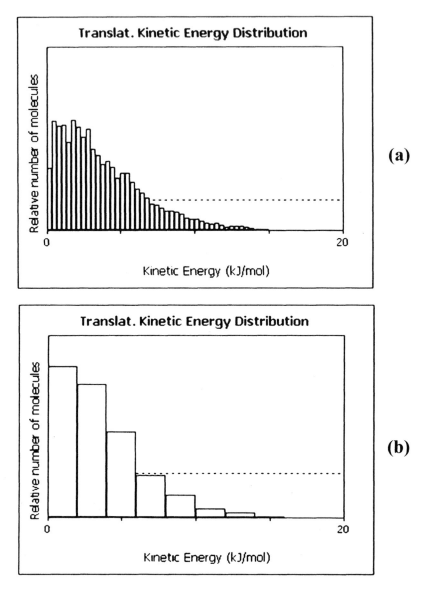

Figure 3. Kinetic energy distribution calculated with Odyssey for a sample of carbon monoxide. The bin width of the histogram is either smaller (a) or larger (b) than that in Figure 1.

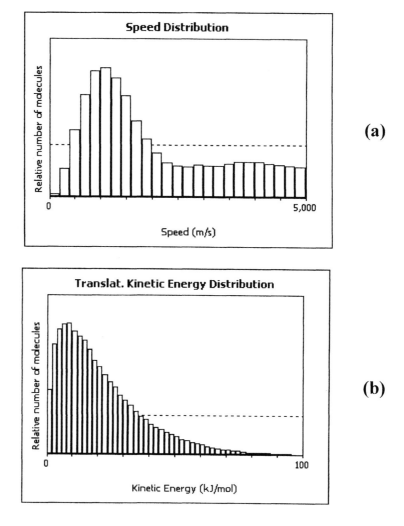

Figure 4. Speed distribution (a) and kinetic energy distribution (b) as calculated with Odyssey for a gas mixture at ~1,500K and ~250 atm that contains 50 molecules of CO and 50 molecules of H_2.

molecules each of carbon monoxide and hydrogen ($T \sim 1500$ K, $P \sim 250$ atm). It can be seen that only the Maxwell-Boltzmann distribution of speeds (a), but not the distribution of kinetic energies (b) is bimodal. While textbooks will generally not apply kinetic theory to gas mixtures, this is a straightforward extension of the simulation approach that illustrates the equipartitioning of energy from a somewhat different angle. In short, the ability of open-ended simulations to go further—sometimes perceived as a threat to a tight teaching schedule—offers an excellent opportunity to apply the concepts learned.

Figure 5 shows the total energy as a function of temperature for a sample of liquid water ($N = 80$), studied in two separate Odyssey experiments. Both plots were recorded by changing the temperature of *one* sample of water, but the length of the individual simulations at each temperature is longer in the first experiment (a) than in the second experiment (b). The goal of the experiments is to determine the heat capacity at constant volume which can be extracted from the slope of the two curves. (Although linear regression lines are calculated, Odyssey does not communicate slopes and intersects and thus forces students to actively go through the corresponding cognitive steps.)

The implicit error bars for the two experiments differ due to the different lengths of the simulations. Interestingly, there is not only the issue of sufficient sampling. If the waiting time between measurements is not long enough, hysteresis from the previous data point (i.e., the previous temperature) also contributes to a less accurate measurement.

Results from a series of Odyssey simulations of non-ideal gases are shown in Figure 6. The compression factor PV/nRT is plotted as a function of the pressure for two systems. The first system is a mixture of hydrogen and helium ($T \sim 120$ K; 90 and 10 molecules, respectively) as it might be encountered in the atmosphere of Jupiter. The second system is pure gaseous ammonia ($T \sim 298$ K; 50 molecules).

The simulations suggest that negative deviations from ideality (corresponding to attractive interactions) are possible for ammonia, but not for a sample of gas consisting mostly of hydrogen. This is in fact what is observed experimentally. Atomistic simulations are by no means guaranteed to reproduce experimental data as well as in this case. Conceptually, however, such simulations tend to be akin to real experiments even when they "fail."

Conclusions

Applicable to topics from the thermodynamics part of the standard curriculum, atomistic simulations allow students to learn physical chemistry with the aid of a laboratory-like tool. The fact that such simulations are not sanitized so as to remove the inherent ambiguity and complexity of real experiments is a major advantage, rather than disadvantage. From a pedagogical standpoint, imperfect data are not a nuisance, but in fact desirable.

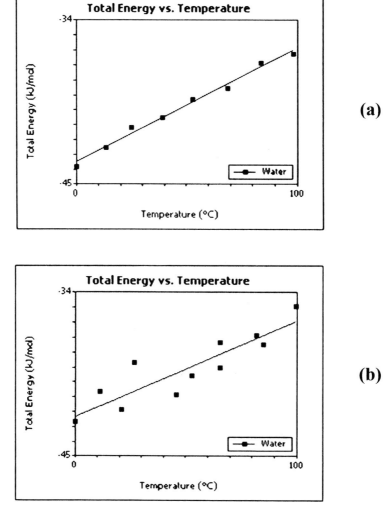

Figure 5. Energy as a function of temperature of a 80 molecule sample of liquid water. The runs underlying each datapoint are longer in (a) than in (b).

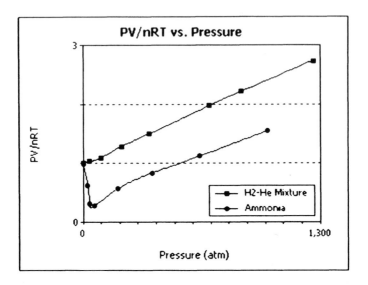

*Figure 6. Non-ideal gas behavior of a mixture of hydrogen and helium
(T ~120K; 90 and 10 molecules, respectively) and a 50 molecule sample of
ammonia (T ~298 K), both simulated with Odyssey.*

Compelling visualization is a defining characteristic of modern software. The three-dimensional models of atomistic simulations in particular are very effective in nurturing an intuitive sense for molecular systems. Operating an apparently "intelligent," interactive, and very visual piece of software also offers students a steady stream of motivational experiences—a benefit appreciated by any instructor.

Interestingly, atomistic simulations are among the very few software applications that actually *make use* of the astounding resource that is available, without any extra investment, to any student or instructor with a computer. Technology has advanced to the point that now even the most inexpensive hardware runs at Gigahertz speeds. While this enormous computational power is irrelevant for many browser plug-ins and other applications used in teaching, it is crucial for the kind of simulations described here. (Video gaming is the other main application to exploit available resources.)

Lastly, the ability to model systems from the "real" world of chemistry means that atomistic simulations are a perfect complement to the abstract models (harmonic oscillator, hydrogen atom, ideal gas, etc.) that are the traditional focus of physical chemistry textbooks. Students are left with a more realistic idea, and greater appreciation, of science.

References

1. *Webster's Encyclopedic Unabridged Dictionary of the English Language*; Random House: New York, NY, 1996.
2. Bourque, D.R.; Carlson, G.R. *J. Chem. Educ.* **1987**, *64*, 232.
3. Geban, O.; Askar, P.; Ozkan, I. *J. Educ. Res.* **1992**, *86*, 5.
4. De Jong, T.; Van Joolingen, W.R. *Rev. Educ. Res.* **1998**, *68*, 179.
5. Barnea, N.; Dori, Y.J. *J. Sci. Educ. Tech.* **1999**, *8*, 257.
6. *Wikipidia*; en.wikipedia.org/wiki/Simulations.
7. *Mathematica*; Wolfram Research, Inc., Champaign, IL.
8. *Maple*; MapleSoft, Waterloo, Ontario, Canada.
9. *MathCad*; Mathsoft Engineering & Education, Inc., Cambridge, MA.
10. *Interactive Physics*; Design Simulation Technologies, Inc., Canton, MI.
11. *Gaussian*; Gaussian, Inc., Wallingford, CT.
12. *Spartan*; Wavefunction, Inc., Irvine, CA.
13. *CAChe/MOPAC*; Fujitsu, Inc., Beaverton, OR.
14. *HyperChem*; Hypercube, Inc., Gainesville, FL.
15. *PCModel*; Serena Software, Inc., Bloomington, IN.
16. *AMPAC*; Semichem, Inc., Kansas City, MO.
17. *GAMESS*; Iowa State University.
18. *Odyssey*; Wavefunction, Inc., Irvine, CA.
19. *CHARMM*; Harvard University.
20. *AMBER*; Scripps Research Institute.
21. *NAMD/VMD*; University of Illinois.
22. *LAMMPS*; Sandia National Laboratory, Albuquerque, NM.
23. *PINY*; Univ of Pennsylvania, Indiana University, New York University.
24. *Gromos*; Biomos b.v., Zürich.
25. *Gromacs*; www.gromacs.org.
26. *Cerius2*; Accelrys, Inc., San Diego, CA.
27. *Sybyl*; Tripos, Inc., St. Louis, MO.
28. *Atomic Microscope*; Stark Design, Inc., Hoboken, NJ.
29. *Water/VMDL*; Boston University.
30. *Moscito*; University of Dortmund, Germany.
31. *DL_POLY*; Daresbury Laboratory, Cheshire, UK.
32. *Virtual Substance*; University of North Carolina. Also see contribution by C.D. Bruce and J.M. Papanikolas elsewhere in this book.

Chapter 13

Introduction of a Computational Laboratory into the Physical Chemistry Curriculum

Roseanne J. Sension

Departments of Chemistry and Physics, University of Michigan,
Ann Arbor, MI 48109–1055

Computational methods are of increasing importance in the chemical sciences. This paper describes a computational chemistry laboratory course that has been developed and implemented at the University of Michigan as part of the core physical chemistry curriculum. This laboratory course introduces students to the principle methods of computational chemistry and uses these methods to explore and visualize simple chemical problems.

Computational methods have long been important within the physical chemistry community. However, a quick survey of the literature in any of the major chemistry journals will highlight the impact that computational methods are now having in all areas of modern chemistry, from organic synthesis, to biochemistry, to the development of materials, in addition to the traditional areas of physical chemistry. A recent search of American Chemical Society (ACS) Journals for articles published over a four year period returned over 10,000 articles using molecular dynamics simulations. Approximately 40% of the articles were published in the Journal of Physical Chemistry A or B. The other citations were scattered throughout the remaining journals published by the ACS. Computational methods have become ubiquitous in large part because they enhance the microscopic, atomistic, and dynamic understanding of molecular systems. Scientists are able to develop a level of insight and intuition not easily obtained from experiments or equations alone.

The power of computational methods to enhance student understanding of chemical phenomena is, if not ignored, severely underutilized. Future scientists in all areas of chemistry will be expected to appreciate and evaluate the results of simulations and other calculations presented in lectures, seminars, and research articles. Many of them will be required to use sophisticated computational methods in the course of their own research. Today's students should learn how to evaluate the validity of computationally generated results and to make sound choices between the different methods available to address a given problem. Computational methods are not well integrated into most of the undergraduate curriculum, although implementation in the physical chemistry curriculum has been reported (1-5). The approach outlined here builds on many of these earlier ideas while emphasizing collaboration and exploration.

Outline of the Course

The Computational Chemistry laboratory course developed at the University of Michigan is designed to educate students in computational methods within the context of the undergraduate physical chemistry curriculum. The course is a one-credit course required of all chemistry majors, accompanying two semester long lecture courses in Physical Chemistry and a three-credit hour Physical/Analytical laboratory course. The material developed in this laboratory introduces chemistry majors to methods in modern computational chemistry and molecular modeling, and uses these methods to draw connections between the equations presented in lecture, results obtained in physical measurements, and "real" chemical phenomena. Students come out of the course with an appreciation for the methods of modern molecular modeling and an improved insight and intuition for chemical phenomena. The cornerstone of this curriculum is a series of hands-on laboratories. The students learn by doing in a collaborative, exploratory environment. Thus the laboratory units are intended to meet two separate but complementary goals.

1. Each laboratory unit introduces and develops the fundamentals of one of the important computational methods commonly used in modern chemical research. This includes an appreciation for the pitfalls that may arise in any calculation and the ways to avoid these problems.
2. Each laboratory is designed to help students investigate, visualize and explore a chemical problem, developing an insight and intuition not easily developed from equations alone.

The prerequisites for the course include two years of chemistry, including organic, analytical, and an introductory inorganic chemistry course, one year of calculus-based physics, three terms of calculus, and introduction to differential equations usually taken concurrently. Most students take the computational laboratory concurrent with the physical chemistry lecture course covering quantum chemistry and molecular spectroscopy. The course does not require or presume any prior programming experience or any prior experience with computational chemistry software.

The credit load for the computational chemistry laboratory course requires that the average student should be able to complete almost all of the work required for the course within the time constraint of one four-hour laboratory period per week. This constraint limits the material covered in the course. Four principal computational methods have been identified as being of primary importance in the practice of chemistry and thus in the education of chemistry students: (1) Monte Carlo Methods, (2) Molecular Mechanics Methods, (3) Molecular Dynamics Simulations, and (4) Quantum Chemical Calculations. Clearly, other important topics could be added when time permits. These four methods are developed as separate units, in each case beginning with the fundamental principles including simple programming and visualization, and building to the sophisticated application of the technique to a chemical problem.

Each unit is introduced by a sixty to ninety-minute lecture providing an overview of the method, some necessary background information not otherwise covered in the curriculum, and an outline of the goals of the experiments and exercises. Thus the total lecture time over the course of the semester is four or five hours. The course is designed to facilitate hands-on exploration and active learning as much as possible. In this context the course cannot and does not provide comprehensive coverage of computational chemistry.

One Example: Molecular Dynamics Calculations

The molecular dynamics unit provides a good example with which to outline the basic approach. One of the most powerful applications of modern computational methods arises from their usefulness in visualizing dynamic molecular processes. Small molecules, solutions, and, more importantly, macromolecules are not static entities. A protein crystal structure or a model of a DNA helix actually provides relatively little information and insight into function as function is an intrinsically dynamic property. In this unit students are led through the basics of a molecular dynamics calculation, the implementation of methods integrating Newton's equations, the visualization of atomic motion controlled by potential energy functions or molecular force fields and onto the modeling and visualization of more complex systems.

A Simple One-Dimensional Trajectory

A very simple implementation of a molecular dynamics trajectory calculation is achieved by using a velocity Verlet algorithm to calculate the motion of a diatomic vibrator *(6)*. In this method, given an initial position, $x(0)$, and initial velocity, $v(0)$, the trajectories are calculated iteratively from:

$$x(t + \delta t) = x(t) + \delta t \cdot v(t) + \frac{1}{2} \delta t^2 \cdot a(t)$$

$$v(t + \delta t) = v(t) + \frac{1}{2} \delta t \cdot (a(t) + a(t + \delta t))$$

The acceleration *a(t)* is calculated from force field, *f(x,t)* as:

$$a(t) = \frac{f(x(t))}{m} = -\frac{1}{m} \left[\frac{dV}{dx} \right]_{x(t)}$$

For an harmonic oscillator the potential energy function, *V(x)* is:

$$V(x) = \tfrac{1}{2} k x^2$$

and the acceleration is:

$$a(t) = -kx / \mu$$

where μ is the reduced mass of the diatomic and *k* is the bond's force constant.

The harmonic oscillator trajectory calculation is easily programmed in a manner transparent to most chemistry students – including those with little or no programming background. Although many different mathematical packages could be used, we have opted to use the MathCAD package. Given molecular parameters (equilibrium position r_e in Å, force constant k in N/m, reduced mass μ in ng/molecule), program parameters (timestep δt in fs and number of steps N_{TOT}), and initial conditions (velocity, vel, in Å/fs, and initial position x in Å) a simple eight line program MD_1 is used to calculate the trajectory. The MathCAD worksheet programming this calculation is included on the next page, to illustrate how this code is programmed in the MathCAD environment. The program MD_1 will output a matrix with two columns containing the displacement from equilibrium and the velocity respectively at each time step in the simulation. The program is sufficiently transparent for students to modify it to calculate trajectories for the more physically realistic Morse oscillator potential for the same diatomic molecule:

$$V(x) = D_e \left[1 - e^{-\beta x} \right]^2$$

This modification requires calculating the derivative of the Morse Oscillator

Harmonic Oscillator Simulation:

In order to make the example concrete rather than abstract lets consider the vibration of a diatomic iodine molecule. First define the parameters of diatomic iodine: the equilibrium separation in Angstroms = 10^{-10} m, the force constant, k, in N m^{-1}, and the reduced mass in ng.

$$r_e := 2.67 \qquad k := 171 \qquad m_1 := 126.904 \qquad m_2 := 126.904$$

$$N_A \equiv 6.0221367 \cdot 10^{23} \qquad \mu := \frac{m_1 \cdot m_2}{m_1 + m_2} \cdot \frac{10^9}{N_A}$$

$$\delta t := 2 \qquad N_{TOT} := 400 \qquad x := 2.3935 \qquad vel := 0$$

$$
MD\_1 := \begin{vmatrix}
r_0 \leftarrow x - r_e \\[4pt]
v_0 \leftarrow vel \\[4pt]
a \leftarrow \dfrac{-k \cdot r_0}{\mu} \cdot 10^{-18} \\[6pt]
\text{for } i \in 1 .. N_{TOT} \\[4pt]
\qquad \begin{vmatrix}
r_i \leftarrow r_{i-1} + \delta t \cdot v_{i-1} + \dfrac{1}{2} \cdot \delta t^2 \cdot a \\[8pt]
v_i \leftarrow v_{i-1} + \dfrac{1}{2} \cdot \delta t \cdot a + \dfrac{1}{2} \cdot \delta t \cdot \dfrac{-k \cdot r_i}{\mu} \cdot 10^{-18} \\[8pt]
a \leftarrow \dfrac{-k \cdot r_i}{\mu} \cdot 10^{-18} \\[6pt]
\text{continue}
\end{vmatrix} \\[6pt]
MD \leftarrow \text{augment}(r, v) \\[4pt]
MD
\end{vmatrix}
$$

Note that the 10^{-18} in the definition of acceleration is a unit conversion. The force constant k is in units of N m^{-1}, with 1 N = 1 kg m/s^2, therefore, k is in units of kg/s^2. This is converted to ng/fs^2:

$$1 \text{ kg} = 10^3 \text{ g} = 10^{3+9} \text{ ng} = 10^{12} \text{ ng}$$
$$1 \text{ s} = 10^{15} \text{ fs}$$

Thus kg/s^2 = 10^{12} ng/10^{30} fs^2 = 10^{-18} ng/fs^2.

The following line calculates the trajectory.

$$\text{Trajectory} := MD\_1$$

Figure 1. Mathcad worksheet calculating the harmonic oscillator trajectory with parameters appropriate for the iodine molecule.

potential, putting the derivative into the calculation of the acceleration, and making the necessary changes to get the units right.

The initial exploration in this unit requires the students to compare the trajectories calculated for several different energies for both Morse oscillator and harmonic oscillator approximations of a specific diatomic molecule. Each pair of students is given parameters for a different molecule. The students explore the influence of initial conditions and of the parameters of the potential on the vibrational motion. The differences are visualized in several ways. The velocity and position as a function of time are plotted in Figure 2 for an energy approximately 50% of the Morse Oscillator dissociation energy. The potential, kinetic and total energy as a function of time are plotted for the same parameters in Figure 3.

The trajectories plotted in Figure 2 highlight the differences between the two potential energy functions and provide insight into the nature of harmonic and anharmonic motions at the inner and outer turning points. In particular the difference in time spent at the inner and outer turning points for the anharmonic potential correlates with the "sawtooth" periodicity in the velocity. Plots of the kinetic, potential, and total energy also provide insight into the distinctions between harmonic and anharmonic motion and the implications of the potential energy function (Figure 3).

Although the 2-d plots contain a great deal of information, the most useful visualization tool for most students is the animation or movie. The students prepare animations of the harmonic and anharmonic motion for direct

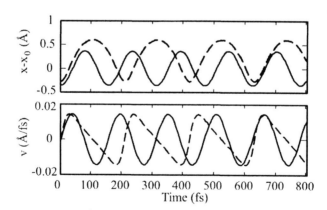

Figure 2. Internuclear separation (top panel) and velocity (bottom panel) as a function of time for a Morse (dashed line) and harmonic (solid line) oscillator having the same total energy, ca. 50% of the dissociation energy of the Morse potential. Note the rapid change in velocity at the inner turning point and slow change in velocity at the outer turning point for an anharmonic oscillator. This reflects the slope of the potential in each case.

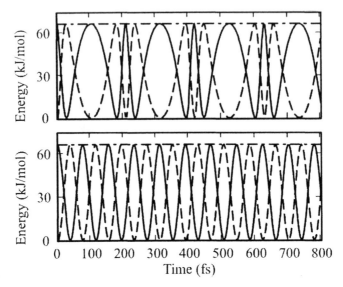

Figure 3. Potential energy (solid line), kinetic energy (dashed line), and total energy (dot-dashed line), for an anharmonic (top panel) and harmonic (bottom panel) oscillator with the same ground state vibrational frequency. The parameters are the same as in Figure 1.

comparison with the 2-d plots in Figure 2. These animations may be prepared for the potential by plotting the bond length as a function of time, or by plotting the positions of the two atoms as a function of time. Snapshots for these animations are illustrated in Figure 4. Comparison of the motion – slow at the outer turning point and rapid at the inner turning point – builds insight into the interpretation of graphs such as those plotted in Figures 2 and 3, and into the implications of changes in slope and curvature of potential energy functions.

In addition such plots provide a good chance to explore the influence of the details of the calculation, such as the time step, on the ability of the algorithm to produce reliable results. One of the goals of the course is to provide students with an awareness of and watchfulness for the signs of computational failure. One example of failure is plotted in Figure 5. Variation of the time step leads to a situation where the total energy of the oscillator changes as a function of time – a clearly non-physical result. As the iterative solution to Newton's equations proceeds the system gains energy and eventually dissociates

Extension to Many Dimensions

Simple one-dimensional trajectories are important tools for developing an understanding of the strengths and weaknesses of molecular dynamics methods.

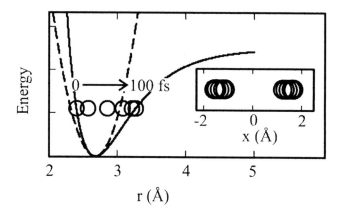

Figure 4. Still shots from movies of the motion of an anharmonic oscillator with a total energy approximately 50% of the dissociation energy. The system starts at the inner turning point and travels to the outer turning point. The students make animations of these plots to compare the observed motion with the calculations plotted in Figures 2.

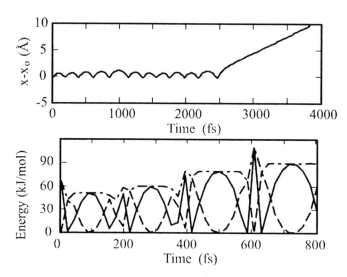

Figure 5. Calculation for an anharmonic Morse oscillator, as in Fig. 3 with a larger time step. This causes the system to gain total energy and eventually dissociate. This is one example of a failed simulation.

Extension to many dimensions provides insight into more sophisticated aspects of the method and into the nature of molecular interactions. In the second stage of this unit, the students perform molecular dynamics simulations of 3-D van der Waals clusters of 125 atoms (or molecules). The interactions between atoms are modeled using the Lennard-Jones potentials with tabulated parameters. Only pairwise interactions are included in the force field. This potential is physically realistic and permits straightforward programming in the Mathcad environment. The entire program is approximately 50 lines of code, with about half simply setting the initial parameters. Thus the method of calculation is transparent to the student.

The output of the calculation is analyzed by making movies of the cluster dynamics, calculating radial distribution functions and velocity-velocity autocorrelation functions and by investigating specific trajectory paths describing the motion of individual atoms in the system. The pairwise radial distribution function, $g(r)$, is a useful way to describe the structure of a system. The radial distribution function is defined such that $g(r)dr$ is the probability that a particle will be found at a distance between r and $r+dr$ from another particle. This distribution function provides a measure of both short-range and long-range order. The function $g(r)$ is calculated by dividing the distance range into bins and counting the number of atoms falling within each bin for all atoms. The program *dist* is used to calculate $g(r)$. In this program N_{BIN} is the number of bins, N_{START} is the time step at which to begin calculating the radial distribution function. The calculation should use only time steps after the initial equilibration is complete. N_{FINISH} is the last time step to include in the average. Normally this will be the last step of the simulation, but a smaller range may be used (e.g. to keep the calculation from taking far too long). N_{ATOM} is the number of atoms included in the simulation, and PER_BIN is the size of each bin and x, y, and z are matrices containing the coordinates of each atom at each time point.

$$\text{dist} := \begin{vmatrix} i \leftarrow 1.. N_{BIN} \\ \text{for } n \in N_{START} .. N_{FINISH} \\ \quad \text{for } i \in 1.. N_{ATOM} - 1 \\ \qquad \text{for } j \in i.. N_{ATOM} \\ \qquad \begin{vmatrix} r \leftarrow \sqrt{\left(x_{n,i} - x_{n,j}\right)^2 + \left(y_{n,i} - y_{n,j}\right)^2 + \left(z_{n,i} - z_{n,j}\right)^2} \\ k \leftarrow \text{floor}\left(\dfrac{r}{PER\_BIN}\right) \\ A_k \leftarrow A_k + 1 \text{ if } k < N_{BIN} \\ \text{continue} \end{vmatrix} \\ \quad \text{continue} \\ A \end{vmatrix}$$

The output of this program is a vector containing the number of atom-atom distances falling within each bin.

The velocity autocorrelation function, C(t), is given by:

$$C(t) = \frac{1}{N} \sum_{i=1}^{N} \frac{\langle \vec{v}(0) \cdot \vec{v}(t) \rangle}{\langle \vec{v}(0) \cdot \vec{v}(0) \rangle}$$

describing the normalized projection of the velocity at time t on the velocity at time zero averaged over all of the particles in the system. This function is calculated in MathCAD as an average over v(0)·v(t) using a range of different time points as t=0. *(6)*

The calculations of g(r) and C(t) are performed for a variety of temperatures ranging from the very low temperatures where the atoms oscillate around the ground state minimum to temperatures where the average energy is above the dissociation limit and the cluster fragments. In the course of these calculations the students explore both the distinctions between solid-like and liquid-like behavior. Typical radial distribution functions and velocity autocorrelation functions are plotted in Figure 6 for a van der Waals cluster at two different temperatures. Evaluation of the structure in the radial distribution functions allows for discussion of the transition from solid-like to liquid-like behavior. The velocity autocorrelation function leads to insight into diffusion processes and into atomic motion in different systems as a function of temperature.

In the course of these investigations the students also prepare movies of the cluster evolution to better visualize the dynamics represented by the one-dimensional velocity autocorrelation and the radial distribution function plots. One interesting feature apparent in simulations carried with an average energy well above the dissociation energy is the phenomenon of evaporative cooling, where partitioning of some of the energy into the escape of a few atoms leaves behind a bound, discernibly "cooler" cluster. In some simulations the hot cluster fragments into a few smaller clusters with substantially lower internal temperatures. These observations all help to enhance student intuition.

Another useful tool is the comparison of trajectories for individual particles in the cluster. The trajectories of a dozen atoms are compared in the plot below for the same two simulations producing the correlation functions plotted in Figure 7. These plots and others like them allow the student to visualize the distinction between atoms on the surface and atoms buried in the interior of the cluster as a function of temperature.

Extension to More Sophisticated Applications

The first two stages in each unit guide the students through the fundamentals of the computational method and through the application to a

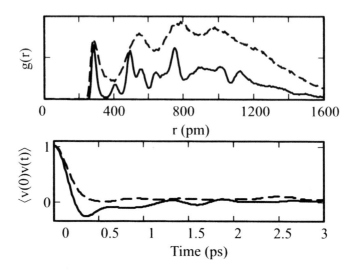

Figure 6. Radial distribution function (top panel) and normalized velocity autocorrelation function (bottom panel) at temperatures equivalent to 20% of the Lennard-Jones well depth(solid lines) and equivalent to 75% of the well depth (dashed lines).

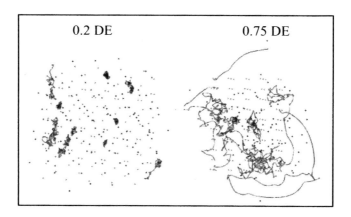

Figure 7. Trajectories for12 atoms in the MD simulations of motion in van der Waals clusters at two different temperatures, with average energies significantly below (left) , or approaching (right) the dissociation energy (DE).

transparent but moderately complex system. In the final stage, the students use a sophisticated computational program to answer a self-contained research question. In the context of the quantum chemistry unit, students have been asked to determine the likelihood of excited state tautomerization (proton transfer) for one of a variety of organic compounds. This is an exploration based on a question originally posed by Moog for use in the Physical Chemistry Curriculum *(7)*. Gentisic Acid and its tautomer provide an example of such a tautomer pair.

In the final exploration of the quantum chemistry unit students use a computational chemistry package (eg. Spartan, Gaussian, CaChe, etc.) to calculate the ground state energies, molecular orbitals, and in some cases the excited state energies, of two proton transfer tautomers. Calculations are performed at several different levels of theory, and use both semi-empirical and ab initio methods. Several different basis sets are compared in the ab initio calculations. The students use the results of these calculations to estimate the likelihood of excited state proton transfer. The calculations require CPU time ranging from a couple of minutes to a couple of hours on the PCs available to the students in the laboratory.

These systems provide a useful example because the calculations often work, but occasionally fail, either by distortion from planarity or by failure to locate a stable minimum for one of the tautomers. Thus, the students learn to consider their results critically with a healthy dose of skepticism, to analyze the success or failure of the calculation, to consider the influence of the choice of method (semi-empirical or ab initio), to consider the influence of the choice of basis set, and to determine the answer to the research question posed.

One of the tautomer energy minimization calculations yielded the structure shown above in Figure 9. In this case the geometry optimization did not find a stable minimum at the desired configuration and the proton was transferred back to the original oxygen. Interestingly in the initial offering of the course, several students performing calculations on a compound suffering from this failure commented that the orbitals and energies were the same for both tautomers, but never noticed that the structures were also the same. One student commented that an O-H bond length of 1.866Å seemed long, but did not realize that the

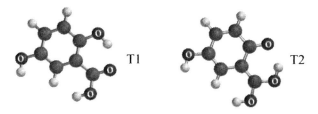

Figure 8. Two tautomers of Gentisic Acid. T1 is the ground state stable isomer.

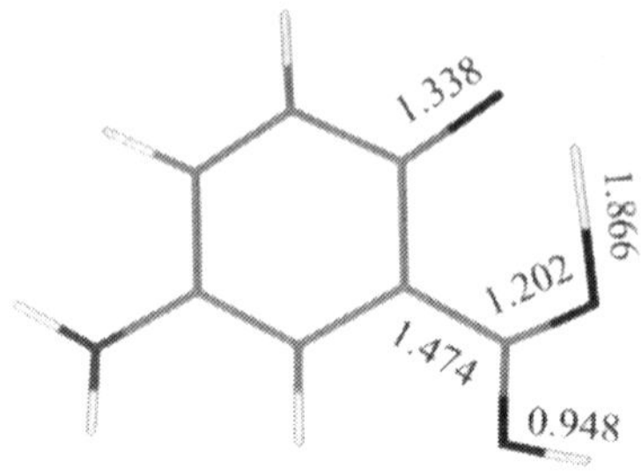

Figure 9. Failed geometry optimization of a proton transfer tautomer.
Black = Oxygen, Dark Grey = Nitrogen, Light Grey = Carbon,
White with outline = Hydrogen.

proton had transferred. In subsequent offerings of the course it was emphasized that calculations can fail, and the students improved markedly in critical thinking and analysis of the results.

Student Evaluation

Student grades in this course are based on written and oral reports. Each student presents two oral reports and turns in two written reports. Students are encouraged to work in groups and talk about the problems and solutions. The requirement that students present their results orally and individually provides each student with the incentive to understand the material. Student understanding is also evaluated through a two-hour final exam. The exam takes examples from the recent literature and requires students to interpret the results shown in the paper and provide a critique highlighting positive and/or negative aspects of the way that a computational method was applied to answer a chemical question.

Summary and Conclusions

Incorporation of computational methods will be a critical feature of the undergraduate education of future chemistry students. Computational methods have become ubiquitous in large part because they enhance the dynamic understanding of molecular systems. Scientists are able to develop a level of insight and intuition not easily obtained from experiments or equations alone. It is vitally important that students understand the fundamental methods, learn to think critically about specific approximations and methods employed, and develop the ability and incentive to evaluate carefully the results obtained from any given calculation. The computational chemistry laboratory course at Michigan teaches students to think about computational methods through a hands-on approach with laboratory units that advance through three stages:

- The first stage begins with simple programs that illustrate the basic principles of the computational method under discussion. The student is then directed to use the method and modify the program to answer a slightly different, slightly more sophisticated, question. This should not be an exercise in programming, but should require the student to understand the mechanics of the method enough to extend the program.

- In the second stage the student uses the computational method in a transparent form to explore a more sophisticated chemical question. Because this is a course in computational methods – not computer programming, these programs are supplied, but are not black boxes. The students have the code before them as they use the programs. The chemical or physical problem posed in this stage should lead the student to a more complete understanding of some chemical phenomenon.

- The third stage involves the use of a sophisticated commercial molecular modeling package to explore a more complex chemical problem in a laboratory setting. The problem posed in this stage should also lead the student to increased chemical and physical insight.

Thus the course meets the complementary goals of increasing student awareness and understanding of methods in computational chemistry and of helping students to investigate, visualize and explore a chemical problem, developing an insight and intuition not easily developed from equations alone. This enhances the students understanding and appreciation of the material developed in the traditional Physical Chemistry lecture courses.

Acknowledgements

As graduate student instructors, Allwyn Cole, David Vodak, Tom Kuntzelman, Andrew Stickrath, and Alex Prociuk all contributed to the development of this course. RJS was supported by the National Science Foundation through the FOCUS Center PHYS- 0114336 and CHE-0078972.

References

1. Jónsson, H. *J. Chem. Educ.* **1995**, *72*, 332-336.
2. Gasyna, Z. L.; Rice, S. R. *J. Chem. Educ.* **1999**, *76*, 1023-1029.

3. Paselk, R. A.; Zoellner, R. W. *J. Chem. Educ.* **2002**, *79*, 1192-1195.
4. Karpen, M. E.; Henderleiter, J; Schaertel, S. A. *J. Chem. Educ.* **2004**, *81*, 475-477.
5. Meija J.; Bisenieks, J. *J. Chem. Educ.* **2004**, *81*, 995-996.
6. Leach, A. R. *Molecular Modeling: Principles and Applications*, 2nd ed.; Prentice Hall: New York, NY, 2001.
7. Moog, R. S. In *Physical Chemistry, Developing a Dynamic Curriculum*; Schwenz, R. W.; Moore, R. J. Eds.; American Chemical Society: Washington, DC, 1993; pp. 280-291.

American Chemical Society Examinations

Chapter 14

The Effects of Physical Chemistry Curriculum Reform on the American Chemical Society DivCHED Physical Chemistry Examinations

Richard W. Schwenz

School of Chemistry, Earth Science, and Physics, University of Northern Colorado, Greeley, CO 80639

The physical chemistry curriculum reform efforts of the last two decades have succeeded in encouraging some revisions in the material in the lecture and in new or modernized exercises for use in the laboratory. More slowly, the mainstream standardized multiple choice examination has also kept pace with the curricular revisions. The content areas represented on the examination have changed with each successive revision of the examination, as have the types of questions asked. The content areas on the examination have become more representative of modern physical chemistry practice, while the items themselves have become more conceptually based.

Physical Chemistry Reform

Physical chemistry, as a separate subdiscipline of chemistry, grew out of the application of the methods of physics to chemical problems. Historically, it distinguished itself from the other subdisciplines of chemistry by its use of mathematics, by the precision with which measurements are performed, and by the emphasis on atomic and molecular processes under examination (1). At the same time as the discipline was developing, a reform of the teaching of chemistry was needed as a discussion of the systematic behavior of reactions was desired to prepare students to deal with the new ways in which material was being discussed.

By the late 20th century, continued calls for the revision of the physical chemistry curriculum were being heard (*2-8*). These calls were for a significant modernization of both the lecture and laboratory curriculum involving an inclusion of modern research topics into the lecture and the laboratory, the deletion or movement of selected material into other courses, and a reduction in the writing requirements for the laboratory. More specifically, the need for experiments and discussion relating to the incorporation of laser and computer technology has intensified with the spread of these devices into all the chemistry subdisciplines. The ACS published a selection of modernized experiments in an earlier volume (*5*).

History of the Examinations Institute and Physical Chemistry Examinations

The Division of Chemical Education of the American Chemical Society created the Examinations Committee (later Examinations Institute) in September 1930 to develop standardized examinations in chemistry (*9*). Initially the Institute began by publishing examinations in general chemistry in 1934. By 2005, the breadth of published materials included examinations for the high school and undergraduate levels in all the subdisciplines of chemistry, booklets concerning test development and administration, study materials for the general chemistry and organic chemistry examination, and small-scale laboratory assessment activities. (ACS Examinations carry secure copyright and as such are released for use rather than published when they are first completed. When an examination is retired – after two new versions of that examination are released – it reaches the point where it is considered published, with the Institute as the copyright holder.) For physical chemistry, currently available examinations include a set of examinations issued during the years 1995-6, and another issued during 1999-2001. A new set of examinations should be completed in 2006. The earlier comprehensive examinations discussed here were issued 1946, 1955, 1964, 1969, 1975, and 1988. Subject area examinations are not discussed here, but are often prepared along with the comprehensive examination.

The physical chemistry examination sets have included three subject area examinations in thermodynamics, dynamics, and quantum mechanics. These examinations would be most useful at institutions on the quarter system where they could be used as final examinations in the respective courses. The final examination in the set is a comprehensive examination covering all three areas of physical chemistry. This comprehensive examination is designed to be used at the conclusion of the year-long course in physical chemistry. In practice however, its most common use is probably as an entrance examination for graduate students. This use raises the question of what material should be on the comprehensive examination because of the nature of its use. Is the comprehensive examination for prospective physical chemistry graduate students? Or all future chemistry graduate students? Or exactly whom? Each of these groups has a different set of content expectations, thus a different set of curricular goals, and a different set of assessment materials is appropriate. Unfortunately, one examination cannot easily address all these issues so that some compromises are necessary. The discussion here addresses only the comprehensive examination, rather than the complete set of examinations.

Why Multiple Choice Examinations?

The use of multiple choice examinations has a long history for the measurement of student achievement, particularly when large groups of students are being administered a common instrument for large scale assessment efforts. There are both advantages and disadvantages in developing this type of instrument. One of the largest advantages lies in the scoring system which is used and in the relatively low cost of scoring. In older instruments, the scoring was often done by hand or on relatively simple computers. Under these conditions it became imperative that there exist a single correct response to each question. This assumption drove the types of questions that could be asked, but also drove the simplicity of the grading system because each item could be scored on a binary basis as either correct or incorrect. Little training is required to score questions and there is little subjectivity as to whether a response is correct or incorrect after an item has been developed and validated. Under ideal conditions, the item responses are all independent of each other and the score on the test can be arrived at from the responses on the individual items. Some research has been done on optimizing the number of responses for any particular item, with three to four responses suggested as optimal (*10, 11*). A typical multiple choice examination should include a variety of question difficulties and types in order to validate that the examination scores are meaningful measures of the student abilities on the tested material, and that the tested material is congruent with both the objectives of the test and the curriculum. More recent suggestions have been made that include giving partial credit for responding with particular incorrect responses, developing linked questions, a wider variety of distractors, and scoring which requires multiple responses to be indicated.

One major perceived disadvantage of multiple choice testing is that only lower order thinking skills are examined. The considerable discussion on this topic has been summarized (*12*). The general opinion is that this disadvantage is a function of the test writers and their items rather than of the test format. A perceived disadvantage is the assumption of a binary scoring system for each item (i.e., no partial credit). This perceived disadvantage is also an advantage because of the lack of subjectivity in the awarding of partial credit, which results in a greater uniformity in scoring. A final relative advantage of the multiple choice format is the larger number of items for students to respond to relative to a constructed response (written out answers) format.

These advantages and disadvantages should be compared with those of a constructed response examination administered to large numbers of students. In such an examination relatively few questions are asked and each is scored on an individual scale, by an individual reader. Consistency in scoring is a major objective of the scoring process across different people scoring the same response. Typically rubrics must be developed which define what determines the assigned point score for a given response. Considerable training for a group of evaluators is involved in achieving a consistency is scoring across examinations. This training often involves considerable expense. The Advanced Placement examination in chemistry (*13*) uses a combination of multiple choice and constructed response items in order to address some of these issues in an examination for high school students at the college general chemistry level. Multiple choice and constructed response scores have been shown to measure slightly different quantities in several situations (*14, 15*).

For these reasons, the Examinations Institute has chosen to continue with the development of sets of examinations in the multiple choice format for a large scale testing program available to a large number of institutions at a reasonable price per examination. The Examinations Institute can consider other means of testing in the future as those means become cost effective.

Types of Multiple Choice Items

Multiple choice examinations are composed of items in a multiple choice format, in which the stem is the part of the question before the multiple choice options. These formats are characterized by the examinee (student) selecting the best response from a set of options. Within this set of parameters, items can be formed in a number of distinct manners. The two forms, which are thought of as the conventional forms, called "complete the sentence" and "give the correct response" do just that. In the first, the stem asks the question in the form of an incomplete sentence and the responses then complete the sentence, for example, "... The properties which must be measured are" for the stem and "pressure and ..." for a response. The second form might ask "What is the original temperature T, in K?" for the stem and a series of numerical values for the responses. These types of questions are commonly used on standardized examinations in a number of fields including chemistry. Most questions on the physical chemistry examinations are in one of these two forms.

Both true-false and extended matching types are considered to be multiple choice formats. When combined with options for "both are true" or "neither are true" they are useful forms of items in some fields. These formats rarely occur on the physical chemistry examination. Another form which is rarely used asks students to fill columns in a matrix format in order to identify which of the sets of crosses of the row and column may be true.

Now that the examination is prepared for publication using desktop publishing software rather than a typewriter, there has been a growth in presentation capabilities so as to allow other forms of questions, such as conventional multiple choice with accompanying graphical material or accompanying tabular material. These two options have expanded the types of questions in an item writer's repertoire considerably. These forms allow the item writer to make the examinee correctly interpret data and/or conceptual information. Another recent form asks the examinee to arrive at an answer from a set of statements. As a simple example, the question might ask "to determine the molar mass of a non-ideal gas, which properties must be measured? I. pressure, II. temperature, III. volume, IV. mass." The responses could be a.) I and IV, and so on. Questions such as these are starting to appear with greater frequency on the physical chemistry examinations. One similar type of question asks to pick the most correct set from a series of choices. For example, "What is the sign of ΔS for the system, the surroundings and the universe for a spontaneous process?", with responses given by combinations of positive and negative signs. A final form is actually two questions linked together. In the first question, the examinee is asked to predict something, and then the second question asks why the response to the first question is picked. Such questions have been field tested in physical chemistry, usually without much success.

Each form of multiple choice item presented is of value to item writers in some field, some are especially valuable for writers in chemistry. The possible forms are expanding as writers' creativity grows and as the modes of presentation improve. Still there are a set of suggestions for writing better items, stated in Haladyna (*16*) which follow.

General Item-Writing Guidelines (Reproduced with permission from reference 16. Copyright 2004 Lawrence Erlbaum Associates.)

Content Guidelines

1. Every item should reflect specific content and a single specific cognitive process, as called for in the test specifications.
2. Base each item on important content to learn; avoid trivial content.
3. Use novel material to measure understanding and the application of knowledge and skills.
4. Keep the content of an item independent from content of other items on the test.
5. Avoid overly specific or overly general items.
6. Avoid opinion-based items.
7. Avoid trick items.

Style and Format Concerns

8. Format items vertically instead of horizontally.
9. Edit items for clarity.
10. Edit items for correct grammar, punctuation, capitalization, and spelling.
11. Simplify vocabulary so that reading comprehension does not interfere with testing the content intended.
12. Minimize reading time. Avoid excessive verbiage.
13. Proofread each item.

Writing the Stem

14. Make directions as clear as possible.
15. Make the stem as brief as possible.
16. Place the main idea of the item in the stem, not in the choices.
17. Avoid irrelevant information.
18. Avoid negative words in the stem.

Writing Options (Responses)

19. Develop as many effective options as you can, but two or three may be sufficient.
20. Vary the location of the right answer according to the number of options. Assign the position of the right answer randomly.
21. Place options in logical or numerical order.
22. Keep options independent; choices should not be overlapping.
23. Keep the options homogenous in content and grammatical structure.
24. Keep the length of the options about the same.
25. None of the above should be used sparingly.
26. Avoid using all of the above.
27. Avoid negative words such as not or except.
28. Avoid options that give clues to the right answers.
29. Make distractors plausible.
30. Use typical errors of students when you write distractors.
31. Use humor if it is compatible with the teacher; avoid humor in a high-stakes test.

Process of Writing the Examination

The process for writing a set of examinations is initiated by the director of the Examinations Institute by their choice of a chair for the development of the suite of examinations in physical chemistry. Following selection of a chair, the chair forms a 10-15 person committee to share in the work of examination development. Members of the committee are selected to balance geography, type and size of institution, interests within physical chemistry, and whether they have served on previous committees. On a side note, it is becoming increasingly difficult to find committee members with interests in thermodynamics. There are term limits for the number of times an individual can serve on the committee; at present that limit is two terms of service. In addition, within the committee as a whole it is necessary to have some members who are outstanding proofreaders, some who understand the typography of chemistry writing, and some who are in touch with how students think. The development of the full suite of examinations for physical chemistry takes three to five years, so participation in the committee entails a long term commitment with work organized in clusters of time, often around the time of an ACS national meeting.

Following selection and approval of the committee members by the Examinations Institute director, the committee begins work on the lengthy process of writing the examination. Typically, the physical chemistry committee meets at the site of each ACS national meeting until work on the examination is completed. Some of the other examination committees may meet at the Biennial Conference on Chemical Education or the ChemEd conference. At the first committee meeting, the committee will typically discuss and make decisions on several items. The first of these items is which examinations will result at the end of the process. Previous committees have opted to write only a comprehensive examination, or a suite of examinations in thermodynamics, dynamics, and quantum mechanics. A second question is how many questions are needed for each examination, and if the multiple choice format is used, how many responses will be used for each question. For example, the committee writing the 2000 set of examinations chose 40 questions with four responses, while the committee for the 2006 set chose 50 questions with four responses. This decision becomes important because it defines the number of questions that need to be written.

Another item relating to the test is how the committee chooses to work. One alternative is to function as a committee of the whole with everyone working on all parts of the examination. A second alternative would be to divide into subcommittees responsible for individual topics. Finally, a number of questions about the question structure need to be answered. One of these involves the use of calculators in taking the examination. The Examinations Institute tries to keep two sets of examinations current so that a committee needs to be aware that an individual examination will likely still be in use of 10 years after its initial issuance. This results in the committee trying to look into the future regarding calculator capabilities, particularly regarding storage and communication capabilities. Lastly, a list of topic areas to be covered is developed for each examination. Among the problematic areas are the placement of statistical mechanics, the inclusion of electrochemistry and phase diagrams, lasers, and modern computational methods.

Each committee meeting will result in homework for the committee members of writing, editing, proofreading, and selection of questions. For example, producing the 50 question examinations resulted from winnowing down over 200 initially drafted questions in each subject area. The most recent committee started work on the dynamics examination first. The process of culling the 200 initial questions to the approximately 100 questions involved question selection, and editing spread over four day-long efforts by the committee. These 100 questions are then formed into two field tests to give to students volunteered by their professors for this purpose. While the field tests were in the hands of students, the effort of editing and question selection continued on the thermodynamics and quantum mechanics questions. One of the areas everyone can help is in the field testing process. Having your students take the field tests improves the statistics used later, and gives additional people an opportunity to look over the field tests for proofreading and content importance purposes.

After a considerable number of students at different institutions have taken the field tests, student response rates for each question are collected and distributed to the committee (17). This data, along with the committee's judgment on subject area distribution, is used to set the final published examinations. This part of the process results in about another 50 questions remaining unselected because they do not discriminate well between poor and good students, or they overlap significantly in content with other chosen questions. Recent committees have tried to choose questions for the finalized examinations so that the average score would be about 50 %. Doing so gives the best results in separating all ability levels of students from each other (18), and has the additional effect of reducing the number of complaints about the difficulty of the examination that occurred occasionally with some earlier examinations. Lastly, the committee proofreads the final form of the examination and examines the norming data.

Question Change with Time

Figure 1 presents data on the number of questions on the physical chemistry comprehensive examination associated with the three subject areas shown as a function of the year of examination publication. Clearly, the relative number of items on thermodynamics, dynamics, and quantum mechanics has changed with time. In order to divide the examination into these three categories, we include statistical mechanics items with thermodynamics, although never more than a few statistical mechanics items have appeared on any individual examination. In addition, we included items related to transport of species within the dynamics portion. This plot gives evidence that the examination content does change with time, albeit slowly. The change has been in the direction towards the areas emphasized by modern physical chemistry research. In looking at the examination items, it is also evident that there has been a change in emphasis within each of the areas. Within thermodynamics for example, the number of electrochemistry items has dramatically decreased over the last 30 years with thermodynamics items becoming focused on applications of the fundamental thermodynamic concepts. The dynamics section has included few items related to transport issues. Those few transport related items now focus on gas phase items including potential energy surfaces rather than on ion transport in solution as previously. The quantum mechanics items have moved from items about the postulates of quantum mechanics and analytical solutions of problems to where they examine spectroscopic application, and a recent move to include items on quantum mechanical methods in structure calculation.

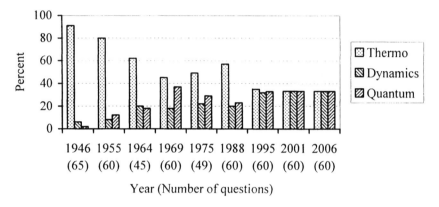

Figure 1. The change of content questions with examination year.

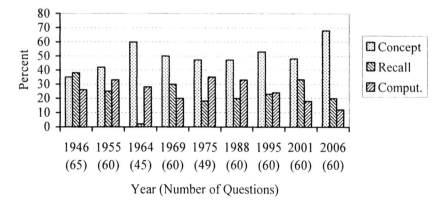

Figure 2. The change in type of question with examination year.

Figure 2 displays information about the types of items on the examinations as a function of time. For this analysis, a conceptual question is one which addresses a fundamental idea without requiring computation beyond that easily done on the examinee's scratch paper, and not simply the restating of facts or definitions. (*19*) Oftentimes, these questions probe student understanding of various representations of macroscopic, symbolic, and microscopic chemical information. (*20*) For example, a question requiring the student to interpret a graph could be a conceptual question. An example non-graphical conceptual question is "At very low concentrations of water in water/ethanol solutions, the Henry's law constant directly relates a) the osmotic pressure of the solution to the concentration of ethanol, b) the equilibrium partial vapor pressure of water to the concentration of water, c) the freezing point depression of the ethanol to the concentration of water, d) the equilibrium partial vapor pressure of ethanol to the concentration of ethanol." A computational question requires the numerical manipulation of a set of numbers to arrive at the correct response. The recall question might ask the student to select an appropriate response from a list of items based on their recall of their class notes or textbook reading. An example recall question might ask, "When a transformation occurs at a constant volume and temperature, the maximum work which can appear in the surroundings is equal to a) $-\Delta A$, b) $-\Delta G$, c) $-\Delta H$, d) $-\Delta S$". Such a classification scheme is not exact because different raters will classify different items into different categories, but it should display trends over time. It is remarkably clear that the percentage of computational items has decreased over time, while the percentage of conceptual items has increased over time, especially with the 2006 examinations. Two suggestions for an explanation for this trend are apparent when the classification is performed. First, the computational items have tended to arise more in the general area of thermodynamics. The decreasing emphasis in thermodynamics with time shown in Figure 1 will directly reduce the number of computational questions. A second effect, especially on the 2006 examination, has been the recognition of calculator capabilities by the committee. A decision was made for the 2006 examination to take calculators out of the hands of the examinees. This decision will make any computational items different, and should reduce their number, thereby increasing the proportion of conceptual items. This observation may be illustrative of a growing trend to emphasize conceptual questions over computational and recall questions in both general and organic chemistry.

Figure 3 presents the data on the changing norms for the examinations over time for the various comprehensive examinations. Data for the 2006 examination is unavailable as the examination is currently in the norm data collection stage. The data is presented as the percent score required to achieve the 35th, 50th, 65th, and 80th percentile score. A percentile score is that required to achieve better than that number of students in a sample of 100 students. Thus a score in the 35th percentile implies that the particular score places the student with a higher score than 34 students in a 100 student sample, and a lower score than 65

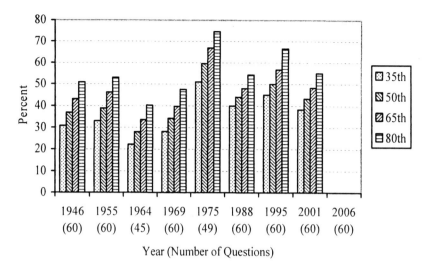

Figure 3. Percentages required for a percentile score as a function of examination year.

students. Figure 3 clearly shows that the percentages required to achieve the same percentile ranking vary with the year the examination became available. Several points require discussion (*21*). First, when a score of 25% or lower is achieved, the examiners worry because no deduction is taken for the number of questions answered incorrectly (since 1973), so a score of 25% should be the score achieved if the student randomly guesses. The 1964 examination was probably too difficult for the students as the 35th percentile score was at, or below, that for random guessing. Second, the data is used to separate the scores for students with different ability levels. Ideally, the bars in Figure 3 would be of much different height for a given examination, resulting from a greater separation of the higher ability students from those of lower ability and greater discrimination between these students.

Upcoming Projects

The physical chemistry community has made several suggestions to the Examinations Institute, as has the community of chemical educators. Following on the success of the study guides to help students prepare for the general and organic chemistry exams, a committee has been formed to develop plans for a study guide for the physical chemistry examinations. As of the writing of this manuscript, the study guide is in draft form, with completion targeted for late spring of 2006. This study guide consists of about 20 chapters with commentary on questions representative of questions for earlier examinations.

Another request from the community of users is for an examination or examinations that could be given at the end of each semester for colleges on the semester system. The committee writing the 2006 set of examinations has responded by writing an examination with a greater number of questions organized into sections. The scoring and norming of the examination will be by section so that professors can select topics for their students to take and then receive norming data for a standardized examination somewhat customized to the material taught in their classroom.

The norming data for all the physical chemistry examinations is now available online at the Examinations Institute web site (*22*). In addition to the norms themselves, it is now possible to enter student response data following taking the examination for comparison with pre-existing data for other schools. Professors will be able to quickly determine how well their students perform relative to their peers at other institutions.

As operations of many types become more developed in an online environment, it would not be unexpected if the format of the examination develops so the examinations could be delivered in an online mode. Using present day technology, it would be relatively straightforward, though time consuming, to adapt the examination to a commercial classroom management system having the capability of multiple choice testing. The difficulty would mostly lie in the complex typography of the items on the examinations and the requirement that security be maintained. Such an online adaptation of the examinations may occur over the next several years. The next big step in the development of how testing occurs will likely be the development of computer adaptive testing environments in which the order of the questions, and even which questions are presented to the students, is a result of the particular correct and incorrect responses a student has given during an earlier portion of the test administration (*23*). The potential advantages of computer adaptive testing include the speed of testing because students are exposed to fewer questions, and the speed with which results are received.

Lastly, in recognition of the growing importance of nanoscale science as an interdisciplinary field with contributions from all the areas of chemistry, and other disciplines the Examinations Institute has begun the process of writing questions developed around nanoscale science concepts, experiments, and methods. We expect that these questions will be additional items which instructors can choose to incorporate into their examinations if those materials are discussed in their courses. These questions are still under development with the committee preparing for publication in 2007.

In conclusion, we have seen that changes have occurred, and will continue to occur, in the format, organization, and content of the physical chemistry examination over time. The multiple choice examination format provides a number of advantages and disadvantages for use in multi-institution testing environments. We expect that modifications will continue to occur in the way the examinations are administered on a longer time scale. We look forward to seeing and experiencing these changes.

Acknowledgements

We appreciate the support of the past and present ACS DivCHED Examinations Institute directors, I. Dwaine Eubanks and Thomas Holme, during the process of writing the suite of examinations, and for their help in providing copies of older examinations and information about those older examinations. I particularly need to thank my colleagues in writing the 2000 and 2006 sets of examinations for their encouragement, their comments, and their hard work on the examinations. They are Lucia Babcock, Juliana Boerio-Goates, Benjamin Degraff, T. Rick Fletcher, Lynn Geiger, Alexander Grushow, David Horner, Peter Kelly, William McCoy, Clyde Metz, Robert Olsen, William Polik, Simeen Sattar, Jodye Selco, Bradley Stone, Marcy Hamby Towns, David Whisnant, Sidney Young, and Theresa Julia Zielinski. Their colleagues have also made the writing of the examinations possible.

References

1. Servos, J.W., *Physical Chemistry from Ostwald to Pauling*. 1st ed, Princeton University Press: Princeton, NJ, 1990.
2. *Content of Undergraduate Physical Chemistry Courses*. 1984, American Chemical Society: Washington DC.
3. *Essays in Physical Chemistry*; Lippincott, W.T., ed.American Chemical Society:Washington, DC,1988
4. Moore, R.J. and Schwenz, R.W. *J. Chem. Educ.* 1992, *69*, 1001.
5. *Physical Chemistry: Developing a Dynamic Curriculum*; Schwenz, R.W. and Moore, R.J., eds.;American Chemical Society:Washington DC,1993
6. http://newtraditions.chem.wisc.edu/PRBACK/pchem.htm
7. Zielinski, T.J. and Schwenz, R.W. *The Chem. Educ.* 2004, *9*, 108.
8. Zielinski, T.J., in *This volume*, M. Ellison and T. Schoolcraft, Editors. 2006, American Chemical Society: Washington, DC.
9. About us, http://www3.uwm.edu/dept/chemexams/about/index.cfm
10. Bruno, J.E. and Dirkzwager, A. *Educ. Psychol. Meas.* 1995, *55*, 959.
11. Haladyna, T.M. and Downing, S.M. *Educ. Psychol. Meas.* 1993, *53*, 999.
12. Osterlind, S.J., *Constructing Test Items: Multiple-Choice, Constructed-Response, Performance, and Other Formats*. 2nd ed. Evaluation in Education and Human Services, Kluwer Academic Publishers: Boston, MA, 1998.

13. AP Central Web Site, http://apcentral.collegeboard.com/article/0,3045,151-165-0-2119,00.html
14. Lukhele, R., Thissen, D., and Wainer, H. *J. Educ. Meas.* **1994**, *31*, 234.
15. Danili, E. and Reid, N. *Chemistry Education Research and Practice* **2005**, *6*, 204.
16. Haladyna, T.M., *Developing and Validating Multiple-Choice Test Items*. 3rd ed, Lawrence Erlbaum Associates: Mahwah, NJ, 2004.
17. Eubanks, I.D. and Eubanks, L.T., *Writing Tests and Interpreting Test Statistics: A Practical Guide*. 1995, ACS DivCHED Examinations Institute: Clemson, SC.
18. Oosterhof, A.C., *Classroom Applications of Educational Measurement*. Merrill Publishing: Columbus, OH, 1990.
19. Nurrenbern, S.C. and Robinson, W.R. *J. Chem. Educ.* **1998**, *75*, 1502.
20. Johnstone, A.H. *J. Comput. Assist. Lear.* **1991**, *7*, 75.
21. Wiersma, W. and Jurs, S.G., *Educational Measurement and Testing*. Allyn and Bacon: Boston, MA, 1990.
22. Main page, http://www3.uwm.edu/dept/chemexams/
23. Wainer, H., *Computerized Adaptive Testing*. 2nd ed, Lawrence Erlbaum Associates: Mahwah, New Jersey, 2000.

Innovative Ways of Teaching

Chapter 15

Walking the Tightrope: Teaching the Timeless Fundamentals in the Context of Modern Physical Chemistry

Michelle M. Francl

Department of Chemistry, Bryn Mawr College, 101 North Merion Avenue, Bryn Mawr, PA 19010

There is a tension in the physical chemistry curriculum between the expanding scope of modern physical chemistry and the timeless fundamentals. Can the curriculum be updated to include modern examples without compromising the basics? A new set of materials for teaching introductory physical chemistry has been developed to address this dichotomy. The materials draw from recent primary research literature to illustrate key principles in multidisciplinary contexts. For example, the kinetics of first-order reactions, illustrated in many current texts by a paper from 1921 reporting the rate of decomposition of gaseous N_2O_5, can be covered instead by following the racemization of amino acids – used in the last decade to date archeological samples.

The Challenge

"The contributions to knowledge in the domain of physical chemistry have increased with such rapidity within recent years that the prospective author of a general textbook finds himself confronted with the vexing problem of what to omit rather than what to include." (*1*)

In 1913 – more than a decade before Schrödinger's groundbreaking work on the wave equation – physical chemistry textbook authors were already lamenting the difficulties engendered by a rapidly changing field. Nearly a century later, the scope of physical chemistry continues to expand, and instructors and authors alike struggle with what to include and what to omit in a one-year required physical chemistry course. Faculty walk a tightrope between covering the timeless fundamentals and exploring the frontiers of research.

Farrington Daniels, who built a strong foundation for physical chemistry instruction with his seminal textbooks, sets out the challenge clearly in his 1931 preface to *Outlines of Theoretical Chemistry*:

"To introduce recent advances without offending old friends who cherish the foundations of a successful past; to keep pace with present tendencies toward the mathematical viewpoint without driving away students who are inadequately prepared; and to sift out the permanent from the trivial are the privileges and responsibilities [of the author]." (*2*)

The tension is clearly perennial. One approach to resolving this tension, at least for many textbook authors, has been the encyclopedic approach, in which modern material is wrapped around the old friends from the successful past. The decisions about what to include in a particular course can then be left to individual instructors. In this chapter I propose an alternative solution to the dilemma, in which recent advances provide the context *within* which the timeless fundamentals are developed, rather than becoming additional material through which both instructor and student must navigate.

The Current Framework

Good research and development demands a knowledge and appreciation of the current state of the field, and pedagogical R&D is no different in this respect. The framework for producing new materials I propose here grows directly from the goals I set out for my physical chemistry course and is conscious of the constraints inherent in the environment in which I teach.

Goals of a First Course in Physical Chemistry

What broad goals should a first, rigorous course in physical chemistry accomplish?

1. To be able to use quantitative, mathematical approaches in the description and analysis of the structure and behavior of matter.
2. To be able to appreciate the strength and limitations of the methods.
3. To be able to apply these methods to other fields, including fields outside of chemistry.
4. To take – and succeed in – a subsequent course in physical chemistry or related subfield.
5. To be able to read the primary physical chemistry literature as a novice.
6. To retain student interest in physical chemistry.

Most physical chemistry instructors probably keep a nearly identical list in the back of their minds. While we may differ slightly in how we rank the various goals, and what weight we give to each, these are the implicit criteria we apply in deciding what materials and pedagogical strategies to use.

Constraints on the System

We tell students, particularly in thermodynamics, that they need to be attentive to the conditions under which particular equations are valid. It is equally critical for instructors to consider the conditions in which we are operating! The key structural pieces of most physical chemistry courses are the text, the problems we assign, and the time instructors and students have to grapple with the reading and working of problems.

Current physical chemistry texts can easily have more than a thousand pages of text and may have as many as 2000 problems to choose from. The selection of problems, not surprisingly given the goals we have for the course, tends to emphasize quantitative results over reflective or critical thinking. Students are not *routinely* asked to consider and comment on the significance of the value they have just calculated, nor to judge how reliable the value might be. Current research often has walk-on role, particularly in the core chapters. It appears in the margins, introductions to chapters, sections on further reading, advanced topics chapters or sections and in the problems. In problems, a reference to the primary literature may be given, but the problem can (and will!) be done without ever reading the paper. These comments are not meant to chide current texts or authors, students would be unable to lift any books trying to meet all these needs simultaneously, but to encourage instructors and authors alike to think about the balance between these factors in any given course or text.

There is also not nearly as much time as you think. As part of the course evaluation process, my institution asks students to estimate the number of hours per week they devote to a course. My physical chemistry students over the last decade (some 200 students) report spending, on average, 10 to 15 hours per week on the lecture part of the course, including class time. So, over the course of a 14-week semester, I can claim roughly 150 to 200 hours of their time. We spend about 40 hours of that in class. What can I expect them to do in the remaining hours? Based on a reading speed of about 100 words per minute for technical material, I can ask them to read about 400 to 450 pages, total, for about 2 hours per week. Assuming that students will need to read the material more than once for comprehension, they are probably not reading nearly as much as you imagine they are, or reading it as carefully as you hope! My own casual ethnographic observations have uncovered that it takes my students roughly 30 minutes a problem (working solo or in a group). This translates to being able to assign 150 to 200 unique problems a semester.

Adding more – readings or problems or topics – likely means that student effort will be more dilute and superficial. There is some evidence that covering more material, particularly in a fashion that intersperses enrichment material throughout the fundamental concepts, results in poorer retention of the fundamentals. (3) Increasing the information density may lead students to retain not only a smaller proportion of the total information presented, but less absolute information than if a smaller amount of material had been covered in the term. Cover less, they may learn more!

Awareness of these constraints suggests that we should avoid adding materials to an already information dense course, and instead focus on the fundamentals of the discipline. If we wish to bring in the current state of the field, it must be taught *through* the fundamentals – not as dessert, but as the main course.

The Problem Begins with the Problems

The problems are the ultimate deliverable in a physical chemistry course. If you can't do the problems, you can't do physical chemistry. Students know this, and focus on the problems, sometimes to the exclusion of reading the text. As a result, I suspect students' primary sense of what the field of physical chemistry comprises, and what it might be useful for, arises directly from the problems assigned. What message do students take home from the problems? While a thorough inventory of the problems available to physical chemistry instructors would be most instructive, the problems collected in Table I, culled from the chapters on chemical kinetics in a number of physical chemistry texts, illustrate the challenges facing the curriculum.

Table I. Selected Problems in Chemical Kinetics

Problem	Source
In the following table are listed specific rate constants k for the decomposition of N_2O_5 at various temperatures. Determine graphically the energy of activation, and find the specific rate constant at 50°C.	(4)
Calculate the number of days required for 10 per cent of a mass of gaseous nitrogen pentoxide to decompose at − 10°C.	(5)
A vessel with gaseous N_2O_5 was immersed at $t = 0$ into a thermostat of 45°C, and the pressure was measured…Find the rate constant k at 45°C for the first-order reaction $N_2O_5 \rightarrow N_2O_4 + \frac{1}{2}O_2$. (Daniels and Johnston, JACS, *43*, 53, 1921.)	(6)
For the gas phase reactions $2N_2O_5 \rightarrow 4NO_2 + O_2$ the rate constant k is 1.73×10^{-5} s^{-1} at 25°C….Calculate the number of N_2O_5 molecules that decompose in 1 s for the conditions [given above].	(7)
The rate constant for the first-order decomposition of N_2O_5 has the value of 4.8×10^{-4} s^{-1}…What will be the pressure, initially 500 Torr, after (a) 10 s…	(8)
The rate constant for the chemical reaction $2N_2O_5 \rightarrow 4NO_2 + O_2$ doubles from 22.5°C to 27.47°C. Determine the activation energy of the reaction.	(9)

All the problems concern the decomposition of N_2O_5, and focus on the fundamentals associated with gas-phase first-order decomposition reactions: calculation of activation energies, pressures as a function of time. This particular example is incredibly ubiquitous. It appears in the problem section and/or the text of nearly every physical chemistry book I have looked in for the last 75 years, beginning with Getman & Daniels in 1931. (10) In many ways, the decomposition of N_2O_5 is an excellent example to illustrate the basics of quantifying first-order kinetic processes. It is clear, clean and draws on real experimental data. But what does it tell students about physical chemistry? In most cases, nothing. Without the context of the literature reference, you could use any other molecule, existing or otherwise, or just call the molecule "A" and students would know as much about the chemistry as they know about N_2O_5 and its significance. In the sporadic instances where the literature reference is given in the problem, students are unlikely to read the paper, and will process at most the date. (*11*) I note that the research was done a decade before *my* parents were born! We might hope that students read this as "the example illustrates something so terribly fundamental we are still reading about it", but my small sample suggests they read it as "there was a date on it?" at best and "old-fashioned and not 'hot' chemistry" at worst. The example continues to be useful in letting students see how the fundamentals "work" and thus well satisfies the first of the broad goals stated above, but fails to move students toward fulfilling others.

Do we really need to fix it?

If it's not broke, don't fix it. As noted in the previous section, the decomposition of N_2O_5 well illustrates a number of timeless foundational principles:

- Simple kinetic motifs
- Graphical analysis of data
- Activation energies
- Rate constants
- Derive integrated rate equations
- Solve rate equations

What does it miss? Students working through this example will not understand the limitations of analyses based on simple reaction motifs. Nor will they see how these concepts can be applied to fields outside of physical chemistry. Many will not be able to develop or interpret other types of graphical analyses based on the equations. And finally, the problem itself is of little or no interest to them.

Can we uncover simple, modern examples that emphasize the same fundamentals, but can catch the missing pieces as well? I would argue that a new example is called for, *not* additional examples, in order to keep the information density at a level that does not compromise student learning.

Context-Rich Materials

Context-rich materials (*12*) embed specific principles drawn from a curriculum into a web of related information. These materials open up for students the "who, what, when, where, and why?" of a concept. The best context-rich materials will offer concrete examples, feature connectivity between disciplines, introduce methodology, and acknowledge the impact solving a problem can have in the broader world. Case studies are one example of context-rich materials that have been successfully used in the teaching of science, but context-rich materials need not be so broad in scope.

There is good evidence that context-rich curricula can attract and retain a diverse group of students, engage students with a wide variety of learning styles, as well as improve student performance. (13-14) Gutwill-Wise compared introductory chemistry courses taught with a context-rich curriculum to those using a conventional pedagogy and text. (13) He found that students in the context-rich sections exhibited a stronger understanding of chemistry, compared to students in traditional sections, and that students emerging from the context-rich track performed at least as well, and in some instances better, than their peers from the traditional track in a subsequent course in organic chemistry. An in-depth and ongoing study of the computer science curriculum at Carnegie-Mellon University (CMU) (14) was prompted by the small number of women enrolling in computer science, as well as the poor retention rate for those who did. CMU reoriented its computer science program to provide a greater emphasis on the context of computing, its relationship to other fields and the overall contributions it can make to society – making it context-rich. From the beginning of CMU's effort in 1995 to 2000, the proportion of women in the computer science program at the institution has increased from 8% to more than 40%.

Context-rich materials and curricula clearly have the potential to make a strong impact on chemistry programs, and in particular physical chemistry. The French National Academy notes "physical chemistry underlies much of modern science and is a motor driving advances in a very wide range of fields. Building on information and concepts from chemistry, physics, mathematics, physical chemistry contributes to and is stimulated by areas as diverse as medicine, molecular biology, biochemistry, molecular engineering, chemical engineering, materials science and earth sciences." (15) Physical chemistry students need to be exposed to a richer set of materials to be well prepared to enter this environment.

Table II gives examples of three problems, ranging from context-free to context-rich. All three cover the same fundamental concepts, but which one looks most appealing to students and would be most interesting to solve? The last certainly excites more curiosity (though savvy students recognize it might be more challenging to solve!).

Strategies for Developing New Materials

Faculty use a variety of materials in teaching physical chemistry, including textbooks, workbooks, and symbolic math workbooks. The best materials – measured against the previously stated goals – are active, current, connected and easy to use. Resources that are *active* require students to engage and reflect on the material at hand. *Current research* should be emphasized. Materials should

Table II. Examples of Problems with Varying Levels of Context

Problem	
Find an expression for the half-life of the third-order reaction $3A \rightarrow B$. (*16*)	Context free
The gas-phase formation of phosgene, $CO + Cl_2 \rightarrow COCl_2$ is 3/2 order with respect to CO. Derive the integrated rate equation for a 3/2 order reaction. Derive the expression for the half-life. (*17*)	Context-ualized
In a paper by Bada, Protsch and Schroeder [*Nature 241*, 394 (1973)], the rate of isomerization of isoleucine in fossilized bone is used as an indication of the average temperature of the sample since it was deposited. The reaction	Context rich

L-isoleucine $\rightarrow$ D-alloisoleucine

iso   allo

produces a non-biological amino acid, D-alloisoleucine, that can be measured using an automatic amino acid analyzer. At 20°C, this first-order reaction has a half-life of 125,000 years and its activation energy is 139.7 kJ/mol. After a very long time, the ratio allo/iso reaches an equilibrium value of 1.38. You may assume that this equilibrium constant is temperature independent.

(a) For a hippopotamus mandible found near a warm spring in South Africa, the allo/iso ratio was found to be 0.42. Assuming that no allo was present initially, calculate the ratio of the concentration of allo now present to the concentration of allo after a very long time (Note: the correct answer is between 0.40 and 0.60).

(b) Radiocarbon dating, which is temperature independent, indicated an age of 38,600 years for the hippo tooth. Using the results of part (a) estimate the half-life for the process.

(c) Calculate the average temperature of this specimen during its residence in the ground. (*18*) demonstrate a clear *connection* to other fields, such as biology, materials science and medicine. Finally, faculty need approaches that can be easily implemented within the context of the traditional three times a week lecture course, and within various course sizes.

New resources are beginning to appear which meet many of these criteria. The latest version of Berry, Rice and Ross (*19*) includes vignettes of key figures in the field that link to problems. Monk (*20*) grounds his physical chemistry text in perspectives drawn from many fields. The POGIL physical chemistry workbooks (*21*) demand full student engagement in the classroom. Alternatively, the collection of symbolic math documents at SymMath (*22*) offers an *a la carte* active approach. Metiu's series of physical chemistry books (*23*) combines the two approaches, embedding the symbolic math work in the text. Here I present another approach, using context-rich materials focused on current research and highlighting the connections to other fields to teach the fundamental concepts.

Culture of Chemistry Materials

The *Culture of Chemistry* materials (*24*) were designed to be context-rich as well as active, current, connected and easy to use. The materials are not meant to teach "library skills", nor are they intended to replace a textbook. They are keyed to fundamental concepts, not adding new topics, but teaching the basics with fresh and modern examples. Each piece stands alone and focuses on a paper from the primary literature. The first set of 6 field-tested modules is:

1. Materials and Nanotechnology: Pulling Gold Nanowires (*Angew. Chem. Int. Ed.* **2003**, *42*, 2251-2253. "Towards 'Mechanochemistry': Mechanically Induced Isomerizations of Thiolate-Gold Clusters")
2. Buckyballs: A Simple Quantum Mechanical Particle on a Sphere Model (*Chem. Phys. Lett.* **1993**, *205*, 200-206. "A particle-on-a-sphere model for C_{60} ")
3. Thermodynamics of Proteins: Calculating the Entropy of a Helix-Coil Transition in a Small Antibacterial Peptide using Statistical Mechanics (*J. Mol. Bio.* **1999**, *294*, 785-794. "Thermodynamics of the α-Helix-Coil Transition of Amphipathic Peptides in a Membrane Environment: Implications for the Peptide-Membrane Binding Equilibrium")
4. Raman Spectroscopy: Detecting forged medieval manuscripts (*Anal. Chem.* **2002**, *74*, 3658-3661. "Analysis of Pigmentary Materials on the Vinland Map and Tartar Relation by Raman Microprobe Spectroscopy ")
5. Exotic Kinetics: Oscillating Reactions in the Troposphere (*J. Phys. Chem. A* **2001**, *105*, 11212-11219. "Steady State Instability and Oscillation in Simplified Models of Tropospheric Chemistry")

6. Archeometry: Using Amino Acid Racemization to Determine the Age of Artifacts (*Science* **1990**, *248*, 60-64. "Dating Pleistocene Archeological Sites by Protein Diagenesis in Ostrich Eggshell ")

Papers from the primary literature are selected carefully. Generally they are short. Reviews are avoided, though the articles need to be readable by a novice. The best papers are cross-disciplinary, in part because of my desire to emphasize the connection to other fields, but also because the interdisciplinary nature of the professional audience tends to promote clearer exposition and limits the use of jargon that can be confusing to the novice.

The pieces include a variety of problems, ranging from basic concepts to critical thinking questions. Examples are given in Table III. The problems are the key to the materials. Students will focus on these, and establishing strong connections to the written material, particularly the article from the literature will make them effective in helping students meet the goals I establish for the course. Students need to read the paper to be able to do the problem! The problems are deliberately "wordy" and presuppose a college reading level. One goal is teach students how to extract the essence of a problem from a lengthy and/or narrative description of a problem. Chemistry in the real world is rarely pre-digested in the way that many chemistry problems are. In assigning problems for students to solve, instructors obviously need to consider the balance between drilling the base skills (where the text of the problems is "bare-bones" and the focus of and approach to the problem is clear) and pushing students to do the initial work of reducing a problem to a clearer statement of what must be done to produce a solution.

Marginalia enrich the pieces, making further connections to other fields, as well as explaining tangential concepts. Examples are given in Table IV. These sections are also meant to expose students to aspects of the history of chemistry and the biographies of chemists. Marginal materials also provide a multiplicity of entry points into the chemistry. Not every student will be excited by every area, but more students may see themselves connected to the field in this way.

Assessment

Faculty

A total of 13 faculty completed anonymous web-based evaluations of the material. Most faculty in the group were tenured (9/13), one-third were at research universities, one-third at liberal arts colleges and the remainder at comprehensive universities. Undergraduate enrollments ranged from 1350 to over 40,000. Chemistry programs ranged in size from 2 majors to more than 70 per year.

Table III. Examples of Problems from Culture of Chemistry Materials

Problem	Objective
In an early paper describing this method [*Science*, **1970**, *170*, 730-732], there is a note that at equilibrium a small excess of D-alloisoleucine is present (at 140°C they report the ratio of alloisoleucine to isoleucine is 1.25), but that for most amino acids the equilibrium constant between the L and D forms is 1. Why should that be? Why do you suppose it is different for isoleucine?	Promote critical thinking skills
Using the expression in question 5 and the parameters given in the paper, compute the forward rate constant at 144°C. What is the value of the rate constant for the reverse reaction at this temperature?	Basic principles
The authors constrained one of the gold atoms to remain in the left-hand plane (see Figure 1 of the paper) as they pull the two planes apart. Identify that gold atom in structure A-I in Figure 2. Could another atom have been reasonably selected? In that case would the results of the authors' simulation be different? In what way?	Promote critical thinking skills

Materials were used both to close a topic, as well as capstone exercises for the term. Nearly all the instructors chose to use selected problems from each module, rather than assign all of the problems in each. One instructor chose to supplement the problems with less complex problems to lead into the exercise. All of the instructors devoted some lecture time to working with students on the exercises, the majority then asking students to complete them outside of class. Some faculty asked students to complete the exercise as groups.

Overall faculty were pleased with the materials – all of them would use the materials again in their teaching. Two-thirds of the faculty felt that students found the modules challenging, while 80% noted that they engaged student interest. The strong connections to modern chemistry were appreciated by two-thirds of the instructors.

The major weaknesses identified were the difficulty of the problems, the need for additional background information for instructors (40%). Faculty (80%) felt that the additional reading sections for students could be eliminated.

Students

Nearly fifty students responded to the anonymous web-based survey. Most (about 80%) were juniors. The majority finds physical chemistry to be an

Table IV. Examples of Marginalia from Culture of Chemistry Materials

Marginal Text
Humans are among the longest-living mammals, with a life span on the order of 100 years. Human ages can be verified by consulting birth and census records. Whales are also apparently very long-lived, but discovering the age of a whale is a somewhat more daunting task and the life spans of most whale species have not been established. The ages of some whales, such as blue whales and fin whales can be determined by counting the layers of earwax in their inner ears. These whales appear to live between 80 and 100 years, similar to humans. Bowhead whales, which live north of the Artic circle, are still hunted in limited numbers by the Inuit. Since the early 1980s, several clearly ancient harpoon heads have been found embedded in modern bowheads. The types of points recovered had not been used since the late 19th century, suggesting the whales were more than 100 years old. Inuit oral histories also supported a long life span for these whales, as multiple generations of hunters described encountering the same whale. The degree of racemization of aspartic acid in the lens of the eye has been used to find the ages in 20 different species, including humans. Ages were obtained for 48 different whales, one that had an apparent age of 211 years, making the bowheads the longest living mammals known!
"The Curta is a precision calculating machine for all arithmetical operations. Curta adds, subtracts, multiplies, divides, square and cube roots... and every other computation arising in science and commerce... Available on a trial basis. Price $125."   From an advertisement in the back pages of *Scientific American* in the 1960s. This is roughly $700 in 2002 dollars – about the same price as *Mathematica*. Curtas sell on e-Bay for thousands of dollars these days.
Does anyone actually use nanowires? Well, Dr. Ock in *Spiderman 2* claimed to use "nanowires" to connect his neural circuitry to a machine circuit. In fact, silicon nanowires have recently been used to build sensors for DNA. The sensors are designed to detect the presence of mutations in a cystic fibrosis gene. [*Nano Letters* **2004**, *4*, 51 -54]

interesting, though stressful course. While they do believe that physical chemistry can provide insights into other fields, they generally do not see how to apply it outside the course. More than three-quarters of the respondents found the materials to be interesting, though almost half found them to be confusing. Almost 90% were pleased to be able to uncover the main idea of the journal articles and a majority found them readable (61%) and interesting (72%). They did find the problems to be challenging (85%). Fifty-two percent found the marginalia interesting, though interestingly, three-quarters of the interested students rated the marginalia as distracting. Just under half the students noted that the marginal materials increased their interest in the exercise. Eighty percent of the students felt the additional reading section could be eliminated as well.

Conclusions: Looking to the Future

It is possible to teach the timeless fundamentals of physical chemistry using modern examples. The key is developing new problems that draw students into, and force them to engage with, the real chemical world. A set of materials that teaches the fundamentals using modern examples has been developed that both students and faculty find to be interesting and challenging. The challenge now to the physical chemistry teaching and research community is to uncover more such examples and develop a variety of strategies for incorporating them into the physical chemistry curriculum.

Acknowledgements

This material is based upon work supported by the National Science Foundation under Grant No. 0340873. Any opinions, findings, and conclusions or recommendations expressed in this material are those of the author and do not necessarily reflect the views of the National Science Foundation.

Many students and faculty used early versions of these materials and their careful critiques and patience are greatly appreciated by the author. The author would also like to thank Eugene J. Miller and the late Lois Cullen Miller for access to their library of classic chemistry texts (and for inculcating in the author an appreciation for the breadth of problems to which chemistry can be applied). Frederick H. Getman, who passed the torch to Farrington Daniels, taught physical chemistry at Bryn Mawr some eighty years before my appointment there. I hope the spirit of excitement in the new developments in the field that infused Getman and Daniels' texts will inspire a new generation of materials.

References

1. *Outlines of Theoretical Chemistry*; Getman, F.H.; Wiley: New York, 1913.
2. *Outlines of Theoretical Chemistry*; Getman, F.H.; Daniels, F.; Wiley: New York, 1931.
3. Russell. I. J.; Hendricson, W. D.; Herbert, R. J. Effects of lecture information density on medical student achievement. *J. Medical Education* **1984**, *59*, 881-889.
4. *Fundamental Principles of Physical Chemistry*; Prutton, C. F.; Maron, S. H.; Macmillan: New York, 1944; p 653. (The text my mother used in physical chemistry.)
5. *Outlines of Physical Chemistry*; Daniels, F.; Wiley: New York, 1948; p 391.
6. *Problems in Physical Chemistry*; Sillen, L. G.; Lange, P. W.; Gabrielson, C. O.; Prentice-Hall: New York, 1952; p 283. (The text my father used in physical chemistry.)
7. *Physical Chemistry*; Levine, I. N.; McGraw-Hill: New York, 1978; p 528. (The text I used in physical chemistry.)
8. *Physical Chemistry*; Atkins, P. W.; W. H. Freeman: New York, 1986; p 710. (The text I used the first time I taught physical chemistry.)
9. *Physical Chemistry: A Molecular Approach*; McQuarrie, D. A.; Simon, J. D.; University Science Books: Sausalito, CA, 1997; p 1178. (The text I use now to teach physical chemistry.)
10. The sample set constitutes the nearly fifty books in my personal collection published between 1900 and 2006.
11. Given the ubiquity of the example, and the dearth of context given in any of the texts I examined, I actually did read the original paper, which notes only that it is of interest to the nitrogen chemists at the Fixed Nitrogen Research Laboratory and as an early (and easily studied) example of first-order decomposition.
12. One definition of context-rich problems is given on page 55 of *Cooperative Group Problem Solving*; Heller, P.; Heller, K.; University of Minnesota: Minneapolis, 1999.
13. Gutwill-Wise, J. P., *J. Chem. Ed.* **2001**, *78*, 684.
14. Margolis, J.; Fisher, A.; Miller, F. "Computing with a Purpose: Gender and Attachment to Computer Science" Carnegie Mellon Project on Gender and Computer Science, working paper http://www-2.cs.cmu.edu/ ~gendergap/purpose.html Margolis, J.; Fisher, A. Unlocking the Clubhouse: Women in Computing, MIT Press, 2001.
15. Pignataro, S. *La Chimica e l'Industria* **1998**, *80*, 1282–1284.
16. *Physical Chemistry: A Modern Introduction*, Dykstra, C. E.; Prentice-Hall: New York, 1997.
17. *Physical Chemistry*, Alberty, R. A.; Sibley, R. J.; Wiley: New York, 1997.
18. *Physical Chemistry: Principles and Applications in Biological Sciences*, I.

Tinoco, Jr., I.; Sauer, K.; Wang, J. C.; Puglisi, J. D.; Prentice-Hall: New York, 2002.

19. *Physical Chemistry*; Berry, R. S.; Rice, S. A.; Ross, J.; Oxford University Press: Oxford, UK, 2000.

20. *Physical Chemistry: A Molecular Approach*; Monk, P. M. S.; Wiley: New York, 2004.

21. *Physical Chemistry: A Guided Inquiry*; Moog, R. S.; Spencer, J. N.; Farrell, J. J.; Houghton Mifflin: Boston, 2004. (The POGIL materials for physical chemistry.)

22. SymMath archive maintained by the Journal of Chemical Education at http://jchemed.chem.wisc.edu/JCEDLib/SymMath/index.html.

23. *Physical Chemistry*; Metiu, H.; Taylor & Francis: New York, 2005.

24. *The Culture of Chemistry*, Francl, M. M.; available at http://www.brynmawr.edu/Acads/Chem/NSFpchem/.

Chapter 16

The Process Oriented Guided Inquiry Learning Approach to Teaching Physical Chemistry

J. N. Spencer and R. S. Moog

Department of Chemistry, Franklin and Marshall College, Lancaster, PA 17604

The pedagogic mainstays of the classrooms in which the authors were educated consisted of lectures, teacher demonstrations of how to use algorithms, student solutions to exercises at the board, and countless homework exercises. Many educators now accept that none of these techniques generally improve student learning or critical thinking skills substantially(1). Indeed, much of the education community now recognizes that listening to an instructor lecture is not an effective way for most students to learn. As Mazur (2) has pointed out, perhaps the survival of these techniques is due to even experienced teachers being misled as to whether students are learning or memorizing algorithms. A more effective learning environment is one in which the students can actively engage, an environment where there is something for students to do. Students tend to stay outside the learning process in a passive classroom or during a laboratory experience in which they follow a set of step-by-step instructions without any requirement for understanding what they are doing. There is no need for much concentration in a passive setting on the part of the student, nor is the student called upon for active engagement (3). A common finding in research on how we learn is that telling is not teaching; an idea cannot be transferred intact from the head of the instructor to the head of the student (4). It is necessary to know what is going on in the student's mind and therefore instructors need to put themselves in a position to be so informed. This new paradigm is put succinctly by Elmore (5):

"Knowledge results only through active participation in its construction. Students teach each other and they teach the instructor by revealing their understanding of the subject."

The Process Oriented Guided Inquiry Learning (POGIL) instructional approach adopts the constructivist view that students construct their own knowledge and that this construction is dependent upon what the student already knows. At the same time, emphasis is placed on the development of important process skills, including higher order thinking skills. In the POGIL paradigm, instructors facilitate learning rather than serving as a source of information, and students work in small self-managed groups on materials specially designed for this approach. In general, these activities guide students to develop the important concepts of the course by using a learning cycle structure. The learning cycle *(6)* is an inquiry-oriented instructional strategy that consists of three general steps. First there is an exploration involving hands-on data, a model, or other information from which the student is guided to the second step, the construction or formation of the concept intended. The third step is an application of what has been learned. Thus, the premises of the POGIL philosophy are that students will learn better when they are actively engaged and thinking in class. They construct knowledge and draw conclusions themselves by analyzing data and discussing ideas. They learn how to work together to understand concepts and solve problems. More information is available from the POGIL website *(7)*.

A POGIL classroom activity begins with a model or information that is to be analyzed or understood. Critical thinking questions then guide students to develop and understand concepts by providing examples of the types of questions chemists would ask about the model or information. The learned concepts are then applied to other problems.

The activities are generally written for a 50-minute class but in some cases less than a period is required and in others more than a period is needed. The activities have frequently been used in 80-minute classroom periods with no transition difficulties from one period to another *(8)*.

The implementation of POGIL principles in the classroom can be accomplished in a variety of ways. A typical implementation in general chemistry has been described previously *(8)*. Additional details about POGIL and its implementation in the classroom and laboratory are available at the website *(7)* and from the POGIL Instructor's Guide *(9)*. As an example, the approach used at Franklin and Marshall College for physical chemistry is briefly described here.

Each class period begins when students enter the classroom and are assigned to groups. Generally there are 3-4 students per group; each student fulfills a role such as manager, recorder of group activities, presenter, or reflector. These roles, which rotate among the students at each class meeting, are defined and explained to the students at the beginning of the semester. The assigned manager picks up a folder that contains any communications the instructor wishes to give the students. The folder also contains a form for the recorder's notes, any quizzes to be returned, and a marked copy of the previous recorder's notes. If the instructor has not done so, the managers assign other group roles. The groups are then asked to answer a focus question for which they often have no firm chemical basis on which to respond. However, this process forces the students to make a prediction based on whatever they do already know. After a brief (2-3 minute) instructor discussion designed to bring into context the day's activity, the groups begin work under the direction of the manager. The instructor moves from group to group listening to group discussions. If an intervention in the group work seems warranted, the instructor may provide some direction to the group. This intervention should not provide students with a direct answer to their questions but should allow the students to discover the concept for themselves. If common difficulties have been observed as the instructor moves about the classroom, these will be addressed later in the period. Another brief (2-3 minute) discussion of the original focus question is often given at the end of the class period, designed to show how the activity had enabled the students to formulate a more meaningful response to that question. Students then place the quiz (assigned the previous period and completed individually outside of class) and the recorder's notes, which contain an assessment of the group's work by the reflector, in the group folder and return the folder to the instructor.

An example of a typical thermodynamics classroom activity is given at the end of this chapter. This activity is done about two-thirds of the way through the semester. Students have had similar activities on real and ideal gases, the first, second and third laws, Gibbs energy, and equilibrium. In these prior activities, the students are guided to develop the important concepts, with the accepted terms describing these ideas often being introduced *after* the concept has been developed. In the example here, the temperature and pressure dependence of phase equilibria for pure phases is explored.

The activity begins with the previously mentioned focus questions, and then some information, in this case the definition of a phase. Next there is a representation (Model 1) of pure liquid water in equilibrium with its vapor. The students begin with the Critical Thinking Questions (CTQs), the concept construction step of the learning cycle. At the end of the activity there is a further table of data for calcite and aragonite followed by CTQ's. Exercises, designed to test the student's understanding, end the activity. The exercises are generally not done during the class period but are assigned as homework. This is the application step of the learning cycle. If a group finishes early and the instructor decides not to have the group proceed to the next activity, a particularly probing exercise may be assigned to the group. If several groups do not finish an activity in the allotted time, the instructor makes a decision as to whether the activity could be finished outside the class and, if so, may make this assignment. Sometimes only minimal discussion is needed to bring closure to the activity and the instructor may choose to do this either by calling on groups who have finished or by guiding the students in a whole class discussion to the completion of the activity.

POGIL Physical Chemistry activities are available from Houghton Mifflin for a two-semester course. The Thermodynamics activities *(10)* include Gases, Thermodynamics, Electrochemistry, Kinetics, and Mathematics for Thermodynamics. The Structure and Bonding activities *(11)* include Atomic and Molecular Energies, Electronic Structure of Atoms, Electronic Structure of Molecules, The Distribution of Energy States and Spectroscopy. As of this writing, activities for a full year of general chemistry *(12, 13)* and a full year of organic chemistry *(14)* have been published; materials for various other courses are currently under development. Up-to-date information can be obtained from the POGIL website *(7)*.

The POGIL approach to laboratory experiences follows this same guided-inquiry framework. The goal of these experiments is to make connections between observations and chemical principles following a learning cycle paradigm. Data is collected and analyzed to support or refute a hypotheses developed by the students. Usually, this is obtained by having each student (or, more often, each group of students) perform *different but related* experiments, so that as a class a wide range of results is obtained from which a trend can be uncovered, or several hypotheses can be tested. The POGIL Project has developed guidelines for these types of experiments, and numerous examples for general chemistry and organic chemistry are available from the POGIL website *(7)*.

At Franklin and Marshall College, this approach has not been used as part of the physical chemistry laboratory component. Rather, the laboratory experience is thought of as an extended "application phase" for many of the concepts developed through the classroom activities. Still, most of the same guiding principles as those used in the classroom are followed — for example, the roughly twelve to fifteen students in a laboratory section are typically divided into about five groups, usually of two or three students. Each group (chosen independently of the classroom groups and kept together for the entire semester) is assigned a different project and given four laboratory periods (four hours, once per week) to develop an appropriate laboratory investigation with guidance provided by the instructor as needed. Each group member submits an individual laboratory report at the end of the development of the assigned study, and the group submits a shortened version of the procedure developed for the project so that other groups may repeat the experiment.

Each group then rotates through all the experiments developed by the class. The course is on thermodynamics and kinetics so the assigned projects cover these topics. A typical set of projects is free energy relationships, activation energy for a reaction, calorimetry, determination of reaction order for a complex reaction, and determination of the thermodynamic parameters for a charge transfer reaction.

For example, Group 1 might be given the project to determine by suitable means the enthalpy, entropy, and free energy change for a charge transfer reaction. Substantially what information the students would be given is that contained in the preceding sentence. The group then goes to the literature to determine the methodology, required instrumentation, and design of the project.

Table I. Laboratory Schedule

Week 1	Check In and Mathematics for Physical Chemistry				
Group	*I*	*II*	*III*	*IV*	*V*
Week 2	CT	K2	K1	C1	GC
Week 3	CT	K2	K1	C1	GC
Week 4	CT	K2	K1	C1	GC
Week 5	CT	K2	K1	C1	GC
Week 6	GC	K1	C1	CT	K2
Week 7	GC	K1	C1	CT	K2
Week 8	K1	C1	GC	K2	CT
Week 9	K1	C1	GC	K2	CT
Week 10	C1	CT	K2	GC	K1
Week 11	C1	CT	K2	GC	K1
Week 12	K2	GC	CT	K1	C1
Week 13	K2	GC	CT	K1	C1

Key: K1 - Oxidation of Alcohols
K2 - Iodination of a Ketone
C1 - Protonization of Imidazole
CT - Charge Transfer
GC - Free Energy Relationships

The instructor provides guidance as the group determines the system to be investigated, the instrumentation that would be most useful, and perhaps some experimental design. The group then develops the experiment and provides instructions for the other groups that will be undertaking this investigation.

After the completion of the four-week projects, each group cycles through all group projects with a two-week period to complete each investigation so that the entire set of experiments developed by all groups is completed. Laboratory reports on the two-week investigations may be individual or group. A schedule of a typical one-semester laboratory sequence is given in Table I.

Students assigned to the charge transfer experiment must first find out what constitutes a charge transfer complex. Then they find examples of charge transfer studies from the literature. They usually quickly see that spectrophotometry is commonly used for such studies and that the temperature will need to be varied. The procedure chosen is to determine the equilibrium constant at more than one temperature and from these data, to calculate the thermodynamic parameters. Students generally have difficulty in selecting a system that has an equilibrium constant that is not too big or too small and to select a solvent. This means they need to do some calculations to determine if a reasonable quantity of product is reasonable quantity of product is produced to allow measurement. Then the appropriate concentrations and the variation in these concentrations must be determined.

By the end of the project, students have developed and carried out an experiment on the thermodynamics of a charge transfer reaction. They also have had the experience of writing guidelines that will permit other groups to carry out the same study.

References

1. *How People Learn*; Bransford, J. D.; Brown, A. L.; Cocking, R. R., Eds.; National Academy Press: Washington, DC, 1999.
2. Mazur, E.; *Peer Instruction: A User's Manual*; Prentice Hall: Upper Saddle River, NJ, 1997.
3. Cracolice, M. S. In *Chemists' Guide to Effective Teaching;* Pienta, N. J.; Cooper, M. M.; Greenbowe, T. J., Eds.; Prentice Hall: Upper Saddle River, NJ, 2005; pp 12-27.
4. Johnstone, A. H. *J. Chem. Educ.* **1997**, *74*, 262-268.
5. Elmore, R. F. In *Education for Judgment*; Christensen, C. R.; Garvin, D. A.; Sweet, A.; Harvard Business School: Boston, MA, 1991; pp ix-xix.
6. Abraham, M. R. In *Chemists' Guide to Effective Teaching;* Pienta, N. J.; Cooper, M. M.; Greenbowe, T. J., Eds.; Prentice Hall: Upper Saddle River, NJ, 2005; pp 41-52.
7. http://www.pogil.org
8. Farrell, J. J.; Moog, R. S.; Spencer, J. N. *J. Chem. Educ.* **1999**, *76*, 570-574.
9. Hanson, D. M. *Instructor's Guide to POGIL*; Pacific Crest: Lisle, IL, 2006.
10. Spencer, J. N.; Moog, R. S.; Farrell, J. J. *Physical Chemistry: A Guided Inquiry. Thermodynamics*; Houghton Mifflin: Boston, MA, 2004.
11. Moog, R. S.; Spencer, J. N.; Farrell, J. J. *Physical Chemistry: A Guided Inquiry. Atoms, Molecules, and Spectroscopy*; Houghton Mifflin: Boston, MA, 2004.
12. Moog, R. S.; Farrell, J. J. *Chemistry: A Guided Inquiry*, 3rd ed.; John Wiley and Sons: Hoboken, NJ, 2006.
13. Hanson, D. M. *Foundations of Chemistry*; Pacific Crest: Lisle, IL, 2006.
14. Straumanis, A. *Organic Chemistry: A Guided Inquiry*; Houghton Mifflin: Boston, MA, 2004.

274

ChemActivity T13

Temperature and Pressure Dependence of Phase Equilibria for Pure Phases

Focus Question: **Does the melting point of ice increase, decrease, or remain constant when the pressure is increased?**

Information

A phase is a region of space in which the intensive properties vary continuously as a function of position. The intensive properties change abruptly across the boundary between phases. For equilibrium between phases, the chemical potential of any species is the same in all phases in which it exists.

Model 1: Two Phases in Equilibrium.

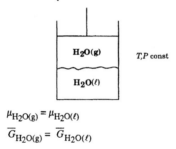

$$\mu_{H_2O(g)} = \mu_{H_2O(\ell)}$$

$$\overline{G}_{H_2O(g)} = \overline{G}_{H_2O(\ell)}$$

Critical Thinking Questions

1. Why can the equilibrium condition for the pure phase equilibria of Model 1 be written as

$$\overline{G}_{H_2O(g)} = \overline{G}_{H_2O(\ell)} \ ?$$

2. Show that $d\overline{G} = \overline{V} dP - \overline{S} dT$ for pure phases where the super bars refer to molar quantities.

Figure 1: Schematic Representation of $\overline{G}$ vs. P for $H_2O(\ell)$ and $H_2O(g)$ at 373 K

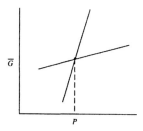

3. In Figure 1:

   a) Label the appropriate solid line "gas" and the appropriate solid line "liquid". Explain your reasoning.

   b) What is the pressure corresponding to the point where the two lines cross (indicated by the dashed line)?

Exercises

1. Use Figure 1 to explain why the boiling point of water is decreased at higher elevations.

2. In Figure 1, the line representing the gas phase should not be straight, expecially over a significant pressure range. Why is this the case? Sketch a graph of $\overline{G}$ vs. P at constant T for a typical ideal gas which shows this curvature, and explain the shape.

Table 1: Standard-State Entropies for H_2O

	$S°$ ($J K^{-1} mol^{-1}$)
$H_2O(s)$	???
$H_2O(l)$	109
$H_2O(g)$	188

$T = 298$ K for liquid and gas. $P = 1$ bar for all phases.

Critical Thinking Questions

4. For a gaseous phase, $\overline{G}(g) = \overline{H}(g) - T\overline{S}(g)$. Provide an equivalent expression for a liquid phase.

5. Consider a plot of $\overline{G}$ vs. T for liquid water. Assume that over the temperature interval considered, $\overline{H}$ and $\overline{S}$, are essentially constant.

   a) Based on your response to CTQ 4, what should be the slope of the line representing the graph of $\overline{G}$ vs. T ?

   b) Based on your response to CTQ 4, what should be the y-intercept (value of $\overline{G}$ when $T = 0$) of the line?

   c) On the figure below, sketch a plot for $\overline{G}$ vs. T for liquid water at 1 bar, and then sketch a plot for $\overline{G}$ vs. T for gaseous water at 1 bar on the same figure. Label each of these lines clearly, including the phase and the pressure.

Figure 2: $\overline{G}$ vs. T for $H_2O(\ell)$ and $H_2O(g)$

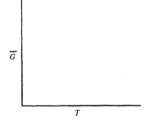

6. Consider the situation in which $H_2O(\ell)$ and $H_2O(g)$ are in equilibrium at a particular T and P.

   a) What condition must $\overline{G}(\ell)$ and $\overline{G}(g)$ meet for the two phases to be at equilibrium?

   b) The two lines that you have drawn on Figure 2 should intersect. If they do not, think carefully about your answer to CTQ 6a and the dependence of $\overline{G}$ on T; then redraw the figure.

   What is the significance of the temperature at which the two lines cross? Explain.

7. We are now interested in adding additional lines to Figure 2—a plot of $\overline{G}$ vs. T for water at 0.5 bar.

   a) How is $\overline{G}$ for a gas affected by changing the pressure at a given temperature? (Hint: Think about how $d\overline{G}$ and dP are related at a given T.)

   b) Based on your answer to CTQ 7a, in which phase is $\overline{G}$ most dramatically affected by changing the pressure? Explain.

d) The graph of $\overline{G}$ vs. T for liquid water at 0.5 bar is essentially identical to the graph for liquid water at 1 bar. Label the line in Figure 2 appropriately and explain this result.

e) What is the significance of the point at which the two lines for P = 0.5 bar intersect?

Information

$$d\overline{G} = \overline{V}dP - \overline{S}\,dT \tag{1}$$

$$H_2O(\ell) \rightleftharpoons H_2O(g) \tag{2}$$

Critical Thinking Questions

8. Consider the situation in which reaction 2 is at equilibrium at some fixed particular T and P.

   a) What is the relationship between $d\overline{G}(\ell)$ and $d\overline{G}(g)$ at equilibrium?

   b) What relationship must exist for the temperatures and pressures of these two phases at equilibrium?

9. Consider reaction 2 at equilibrium.

   a) Use equation 1 to provide an expression relating the volumes, pressures, entropies, and temperatures of the two phases. Use symbols such as $\overline{V}(\ell)$, $\overline{V}(g)$, $\overline{S}(\ell)$, $\overline{S}(g)$, etc. to represent the molar quantities for each phase.

   b) Rearrange your answer to CTQ 9a to obtain

   $$\frac{dP}{dT} = \frac{\Delta_r\overline{S}_{vap}}{\Delta_r\overline{V}} \tag{3}$$

   c) Consider the relationship between $\Delta_r\overline{S}$ and $\Delta_r\overline{H}$ for two phases at equilibrium to show that

   $$\frac{dP}{dT} = \frac{\Delta_r\overline{H}}{T\Delta_r\overline{V}} \tag{4}$$

   d) For the equilibrium $H_2O(\ell) \rightleftharpoons H_2O(g)$ show that

   $$\frac{d\ln P}{dT} = \frac{\Delta_r\overline{H}_{vap}}{RT^2} \tag{5}$$

Table 2: Thermodynamic Parameters for the Two Common Forms of CaCO$_3$, Calcite and Aragonite, at 298 K and 1 atm

	$\Delta_f G_f°$(kJ/mol^{-1})	$\Delta_f H_f°$(kJ/mol^{-1})	$S°$ (JK^{-1}mol^{-1})	V (mL mol^{-1})
Calcite	-1128.8	-1206.9	92.9	36.90
Aragonite	-1127.8	-1207.1	88.7	33.93

Critical Thinking Questions

10. Consider the two common forms of CaCO$_3$ at 298 K and 1 atm.

    a) Which of the two forms of CaCO$_3$ has stronger forces holding the crystal together at this T and P? Explain your reasoning.

    b) Which of the two forms has the most widely spaced and less available energy levels? Explain.

    c) Which of the two common forms of CaCO$_3$ is more stable at this T and P? Explain your reasoning.

Exercises

3. Create a figure analogous to Figure 1 which includes plots for liquid and solid water at 273 K. Label each line carefully, and explain the significance of the temperature at which the two lines intersect.

4. If a wire with weights attached to its ends is placed over a block of ice at its freezing point, the wire will "eat" through the ice. Explain this phenomenon.

5. Integrate the relationship given in CTQ 9d and show how to determine the enthalpy of vaporization of a liquid experimentally.

6. Consider the equilibrium between a pure solid and its vapor and show

$$\frac{d \ln P}{dT} = \frac{\Delta_r \overline{H}_{sub}}{RT^2}$$

where $\Delta_r \overline{H}_{sub}$ is the enthalpy of sublimation.

7. How could the enthalpy of sublimation of a solid be determined experimentally?

8. For the equilibrium between a solid and liquid show that

$$\frac{dP}{dT} = \frac{\Delta_r \overline{H}_{fus}}{T\Delta_r \overline{V}}$$

9. Identify all symbols in Exercise 8. Also show that

$$(P_2 - P_1) = \frac{\Delta_r \overline{H}^{\circ}_{fus}}{\Delta_r \overline{V}} \ln \frac{T_2}{T_1}$$

10. The enthalpy of vaporization of water is 2255 Jg^{-1}. Assume $\Delta_r \overline{H}^{\circ}_{vap}$ to be independent of temperature and pressure and calculate the vapor pressure of water at 80°C.

11. The normal boiling point of benzene is 80.2°C. The vapor pressure at 25°C is 0.13 atm. Find the enthalpy of vaporization of benzene.

12. The vapor pressure of n-pentane is given by $\ln P_{bar} = 8.630 - 2819.7\ T^{-1} + 1.855 \times 10^{-3}\ T$. Derive an expression for $\Delta_r \overline{H}^{\circ}_{vap}$ as a function of temperature and find $\Delta_r \overline{H}^{\circ}_{vap}$ of n-pentane at its normal boiling point 36.1°C.

13. $\Delta_r \overline{H}^{\circ}_{fus}$ of water is 6024 Jmol^{-1} at 273 K. The densities of solid and liquid water are 0.915 and 1.000 g mL^{-1}, respectively. What is the freezing point of water under 100 bar pressure?

14. $\Delta_r \overline{H}^{\circ}_{vap}$ and $\Delta_r \overline{H}^{\circ}_{fus}$ of water are 44.8 KJmol^{-1} and 6024 Jmol^{-1}, respectively at 0°C. The vapor pressure of $H_2O(\ell)$ at 0°C is 4.58 torr. What is the vapor pressure of $H_2O(s)$ at −15°C? State all assumptions.

15. Roughly sketch a plot of $\overline{G}$ vs. T for the two forms of $CaCO_3$. Sketch $\overline{G}$ vs. T for the two forms under very high pressure. At what temperature at 1 bar are aragonite and calcite in equilibrium?

16. At 298 K what pressure is necessary to bring calcite and argonite into equilibrium?

Chapter 17

Teaching Physical Chemistry: Let's Teach Kinetics First

James M. LoBue and Brian P. Koehler

Department of Chemistry, Georgia Southern University, Statesboro, GA 30460

Arguments for the presentation of kinetic theory and chemical kinetics as the first topics taught in the initial physical chemistry course are presented. This presentation allows the first topic in physical chemistry to be mathematically more accessible, to be highly relevant to modern physical chemistry practice, and to provide an opportunity to make valuable conceptual connections to topics in quantum mechanics and thermodynamics. Preliminary results from a recent survey of physical chemistry teaching practice are presented and related to the primary discussion. It was found that few departments of chemistry have adopted this order of topical presentation.

The Philosophy of Teaching

We certainly must ask ourselves what we want our students to learn from a year of physical chemistry. There is far more material in any current textbook than one could hope to cover in any depth. Most would agree that quantum chemistry, thermodynamics, and dynamics must be covered. It is also true that statistical mechanics has increased in importance. Recently we conducted a survey of more than 400 ACS certified chemistry departments in order to investigate trends in physical chemistry teaching (*1*). See Appendix 1 for a summary of results quoted in the text. Of 179 respondents, 122 list statistical mechanics as one of the main topics covered. Beyond the question of what should be taught, we have to ask ourselves what our students' perspective on the subject should be. Unfortunately, more often the following quote seems all too accurate.

> "*Most teachers and students see physical chemistry as an immense body of accumulated knowledge rather than seeing it as avenues that can be taken as we try to make sense of a part of the physical world.*"(*2*)

We submit that the following are important learning objectives to consider:

* Develop an appreciation for the breadth of the subject,
* Foster an understanding for the models most often used, and
* Provide a fundamental understanding that enriches the student's general chemical knowledge.

Improvements in the teaching of physical chemistry have included introduction of modern laboratory experiments (*3*), the use of writing assignments (*4*), and most notably implementation of Guided Inquiry (*5*). A possible re-ordering of topics in the physical chemistry curriculum is also an important reform contribution. Usually, the discussion about the order of topics involves the question of whether quantum chemistry or thermodynamics should be taught first. Moore and Schwenz (*3*) suggested that the teaching of quantum chemistry come first in order to introduce students, early on, to physical chemistry that better reflects current research in the field. In their 1992 paper, they also suggest that the course in the second semester begin with chemical kinetics rather than thermodynamics. In this paper we argue that under the right circumstances the first course in physical chemistry should begin with kinetic theory (including molecular collisions) followed by a presentation of chemical kinetics.

Why We Do It This Way

The Chemistry Department at Georgia Southern has been recognized as one of the top-25 producers of ACS certified majors in the US (*6*). Further, it is the primary contributor to the population of minority bachelors degrees in the physical sciences at Georgia Southern, also with a top-25 US ranking (*7*). Our students come primarily from Georgia and comprise a comparable mixture of both rural and urban students with a significant African-American population.

For the overall health of our program it is critical for us to pay careful attention to the physical chemistry courses which have enrolled 25-30 chemistry majors each year for the past 3 years. The development of our curriculum has always aimed at addressing student learning needs while at the same time inspiring student achievement of high academic standards.

The original motivation for teaching kinetics early in the physical chemistry sequence was, for us, to make sure that enough time was given to the subject to provide proper coverage. Also, because the mathematics is less intimidating, it was a way to give students an early positive experience with the course that is considered the most intimidating for the majority of chemistry majors. At the same time kinetics, immediately following a presentation of gas kinetic theory and elementary collision theory, provides for our students an intuitive picture of how chemical reactions take place.

As a result of this choice of topical order, a number of additional advantages were discovered. If kinetics is developed first, students will very quickly encounter chemical reactions as they did in numerous chapters in their general chemistry course. Unlike either quantum chemistry or thermodynamics, chemical reactions are the focus of a study of chemical kinetics. Further, most students in a physical chemistry course have spent considerable time in organic chemistry studying reaction mechanisms. The goal of a unit on chemical kinetics is to study and test reaction mechanisms. If our students can learn early on that physical chemistry can indeed be comprehensible and relevant to the rest of the chemistry major, it may then be easier for them to believe that the more challenging topics are worth their effort. It is, however, significantly harder to make such a statement if the first physical chemistry topic treated is very challenging and at the same time apparently unconnected or irrelevant (from the students' perspective) to the students' previous experience in chemistry.

Removing the Barriers to Learning

An attempt to soften the mathematical demands in the beginning half of the first physical chemistry course is supported by recent work. Hahn and Polik (8) conclude that mathematical facility is an important factor in a student's success in physical chemistry. Further, Sobzbilir (2), in a study of factors affecting the achievement of students in physical chemistry, concludes that the abstract nature of the content and lack of a connection to the real world impedes learning. As we begin the study of kinetic theory, the most intimidating topic involves the Maxwell-Boltzmann (MB) speed distribution. Fortunately, the concept of a speed distribution is concrete and most likely has been encountered previously in a general chemistry course. Further, many of the derivations in chemical kinetics are short, use only first order, ordinary differential equations, and use calculus in a single dependent variable, thus minimizing the initial mathematical burden on the student. The connection between kinetic theory and chemical kinetics drives home the notion that chemical reactivity depends directly on collision density and on temperature, providing a "first principles" intuitive perspective on rates of reaction. Connecting the tools and concepts of kinetic theory and kinetics to examples from organic chemistry and to biochemistry (through the study of simple enzyme kinetics) gives greater relevance to physical chemistry, further justifying it to our students.

From our own experience teaching the subject we must also add the following observations. A student's lack of self confidence can severely limit their ability to succeed. In addition, the student's preconception of the course (influenced by previous physical chemistry students and sometimes even by their academic advisors) can provide the opportunity for a self-fulfilling prophecy that inhibits learning. The specific mathematical difficulties we encounter have to do with derivations. Students lack the experience to follow derivations and lose interest long before the final result is proven. This has more to do with an inability to do algebra than it does with a lack of background in calculus. One might surmise that a poor background in calculus very likely results from the lack of a facility with "College Algebra." In fact, it is probably this lack of ability (or facility) that is exhibited by a significant fraction of students in each physical chemistry class leading to their lack of confidence which in turn reinforces their preconceptions about the course.

More Arguments

The content of both quantum chemistry and thermodynamics as traditionally presented, although essential to the background of any chemistry major, lacks a strong connection to the content of courses previously taken. This is not to say that physical chemistry textbooks lack connections to a student's previous study of chemistry, but it is clear that the typical student has difficulty seeing those connections.

In thermodynamics, much weight is given to the development of the laws of thermodynamics focusing on calculations of work and heat and presenting or deriving tools used to prove ultimately that ΔS is a state function and that the second law relates mostly to heat engines and refrigerators. The rest of the study of thermodynamics involves the development of tools devoted to the description and investigation of chemical and phase equilibrium. A brief interlude involving thermochemistry and calorimetry is the first connection back to general chemistry content (except for a brief review of the ideal gas law that appears at the beginning of many current textbooks). Chemical equilibrium, which is developed a chapter or two after thermochemistry, is the next topic previously seen in the general chemistry course. These topics are, however, incidental to, as described in most physical chemistry textbooks, the overarching goal of constructing, through considerable derivation, the formal structure of thermodynamics. With chemical equilibrium disguised by thermodynamic formalism, the typical student has difficulty recognizing how it relates back to the equilibrium studied in general chemistry. Of course it is our intent to expand our students' knowledge beyond the chemistry learned in previous courses, but it is far easier to do so from the foundation of previous knowledge. It is the students' attitude that is important here. If our students get lost in the first topic presented, we might not find them again before the end of the course.

The aim in quantum chemistry is, once again, the construction of a formal mathematical structure that is then applied to standard problems, e.g. the particle in a box, the rigid rotor, the harmonic oscillator, the hydrogen atom, etc. With each example, significantly more mathematics is introduced so that the typical student finds it quite difficult to identify any sense of commonality these problems have with one another. Of course this is the nature of the quantum chemistry segment of most physical chemistry courses as described by the most current textbooks. Admittedly, the particle in a box is a one dimensional problem, a second order differential equation, easily solved. Unfortunately, there are few practical chemical applications for this model, and certainly none that might be familiar to the typical student. Thus, the teacher must expend extraordinary effort to motivate students to learn this topic. The harmonic oscillator is covered next, and even though the Schrodinger equation appears to be simple, its solution lies outside of the scope of the course. The focus here is on a new set of eigenfunctions that require significant sophistication to understand. In order to keep the students' interest, the presentation of the two models (harmonic oscillator and particle in a box) must emphasize the commonality of these two topics—oscillatory behavior, even/odd symmetry, zero-point energy, orthogonality, etc. Even so, the student must practice considerable persistence to learn the important quantum chemical principles exemplified in these two problems. And so it goes with the next models presented. As with thermodynamics, the presentation of quantum chemistry focuses as much on the development of a mathematical structure as it does on the applications of the theory. We physical chemists revel in the beauty of this formal structure, but we must ask, how many of our students share our enthusiasm, and more importantly, how can we inspire their enthusiasm? Unfortunately, the development of formal mathematical structure is made to an audience, the majority of whom lack the skills and patience to appreciate those developments.

Physical Chemistry at Georgia Southern

At Georgia Southern we begin the year of physical chemistry with a review of basic, relevant physics concepts including kinetic energy, force, pressure, the ideal gas law, and the units that describe them. This leads into a rather standard development of some of the kinetic theory formulas, especially the collision frequency and collision density. The algebraic derivation of these equations relies on concepts from general physics along with the simple hardsphere model. The primary "take-home" lesson for the student is that chemical reactions, in order for them to take place at all, require collisions. The development of all equations that relate to kinetic theory is taken largely from Laidler, Meiser, and Sanctuary (9) and from Alberty, Silbey, and Bawendi (10). We have greatly expanded on the textbook presentations, making our derivations available to our students on the course web site. This derivation introduces students to the kinetic molecular models, and the derivations that result build on one another and produce several intermediate expressions along the way, including:

$$P \cdot V = \frac{1}{3} \cdot N \cdot m \cdot \overline{u^2},$$
Eqn 1

$$u_{rms} = \sqrt{\overline{u^2}} = \sqrt{\frac{3 \cdot R \cdot T}{M}},$$
Eqn 2

$$\text{and } \overline{E_k} = E_k = \frac{3}{2} \cdot R \cdot T.$$
Eqn 3

Each stopping point gives students a "breather" during which numerical calculations can be carried out (Eqns 2 and 3) or in which a gas law can be discussed (Boyle's Law Eqn 1). This introduction allows for a smooth development of the collision density and collision frequency concepts by getting students used to thinking about molecules from the microscopic standpoint of the kinetic theory. At the same time students begin to realize that physical chemistry is all about proving the validity of the equations used. Because we use the average speed in the derivations instead of the root mean squared speed, we must devote appropriate time to the Maxwell-Boltzmann speed distribution which we present without derivation, instead focusing on explaining the notion of a continuous average. We also solve related problems in class and assign appropriate homework to further support our learning objectives for this topic. It should be noted here that homework is an important part of the course with 8-10 assignments scattered throughout the semester.

Once the collision frequency and density are determined, we focus on example calculations. The capstone for this section is a "ballpark" calculation of the initial rate of reaction at atmospheric pressure for a gas phase chemical reaction at 2700 K. We present the reaction as described by Levine (11) for a simple, bimolecular reaction step.

$$CO(g) \quad + \quad O_2(g) \quad \rightarrow \quad CO_2(g) \quad + \quad O(g)$$

Assuming only that the rate of reaction depends directly on the rate of collisions between CO and O_2, an initial rate can be calculated and then compared with an experimental reaction rate. The calculated value overestimates the reaction rate by more than a factor of 200 providing an opportunity to examine the assumption that all collisions result in a reaction. This example provides us with the opportunity to preview issues to be dealt with in the chemical kinetics chapters that follow, namely the orientation factor and the Boltzmann factor, both of which must be considered if a reliable prediction of reaction rate is to be calculated (at least from the standpoint of hard-sphere collision theory).

We then proceed to a traditional presentation of experimental chemical kinetics. The two important concepts emphasized are, first, bond breaking is endothermic, and second, when simple particles collide and a bond momentarily forms, often a third body must be introduced in order to conserve energy and stabilize the product species. We next present the theory of the rate constant I beginning with the Arrhenius theory and then proceed to hardsphere collision theory, which brings us back to our collision formulas. We next cover Transition State Theory (TST), which at this stage presents some problems as well as opportunities. We use the presentation described in Laidler, et al. (*12*) The problem is that we need to assume that students understand basic principles of equilibrium, that they have some understanding of entropy and enthalpy, and that they have seen the Gibbs energy formulas not yet encountered in the physical chemistry course. We find that the students' knowledge of these concepts and formulas from general chemistry is sufficient. Further, we find that TST is an opportunity to review material our students have already encountered in general chemistry that will again be developed when we cover thermodynamics.

We continue our study of chemical kinetics with a presentation of reaction mechanisms. As time permits, we complete this section of the course with a presentation of one or more of the topics: Lindemann theory, free radical chain mechanism, enzyme kinetics, or surface chemistry. The study of chemical kinetics is unlike both thermodynamics and quantum mechanics in that the overarching goal is not to produce a formal mathematical structure. Instead, techniques are developed to help design, analyze, and interpret experiments and then to connect experimental results to the proposed mechanism. We devote the balance of the semester to a traditional treatment of classical thermodynamics. In Appendix 2 the reader will find a general outline of the course in place of further detailed descriptions.

Because derivations are an important part of the curriculum, it is important to present here our method of dealing with them. We present derivations in electronic format in class and make the same derivations available on the web, allowing students to spend time in class thinking rather than copying. Our method is not specific to our choice of the order of topics, but it is important for the reader to know the details of our implementation. We encourage students to focus on the following points in dealing with derivations:

- The assumptions made at the outset

- The symbols used and their individual meaning
- The approximations invoked along the way and
- The significance of the final result.

If students learn to pay attention to these points, they can take valuable conceptual learning even from derivations they could never carry out by themselves.

The Laboratory

We utilize the laboratory, which is not a separate course, in the process of introducing higher level mathematics. For instance, the first day of laboratory is given to mathematics exercises that review simple integrals and derivatives, and the chain rule. This is also where partial derivatives are introduced using the ideal gas law and the van der Waals equation as object lessons. It is here that we also introduce the triangle derivative rule for partial derivatives, Eqn 4.

$$\left(\frac{\partial Z}{\partial X}\right)_Y \cdot \left(\frac{\partial X}{\partial Y}\right)_Z \cdot \left(\frac{\partial Y}{\partial Z}\right)_X = -1 \qquad \text{Eqn 4}$$

These exercises are easily completed in one, three-hour laboratory period. The capstone to these exercises is a take-home problem introducing the propagation of error formula which we initially present at the end of the first laboratory period. All of these mathematical exercises are presented in Mathcad which allows us to reacquaint our students with the program they learned about in our three-semester-hour Research Methods course usually taken in the sophomore year. We also introduce the symbolic calculus capability of Mathcad that allows simple determination of derivatives and antiderivatives. For the propagation of error exercise we use the formula relating the ideal gas law to the molecular weight of a volatile liquid. We provide the uncertainties in four measured quantities along with values for T, P, V, and mass (of the volatile liquid) and ask for the calculation of the uncertainty in molecular weight. This exercise is handed in the next week of laboratory. Later, during the first round of experiments, students actually determine the molecular weight of a volatile liquid by the Dumas method, and the results they obtained in their take-home problem can be applied to their laboratory report for this experiment.

Insisting that the propagation of error formula be used wherever appropriate in laboratory reports, we find it easier to introduce the total differential once the topic comes up during the thermodynamics portion of the course. The similarity between the propagation of error formula and the total differential provides a more intuitive model for our students. Because our order of topics delays thermodynamics to later in the semester, we have time to emphasize the more concrete example used to determine the uncertainty in a measurement.

Planting Seeds

This ordering of topics also allows a number of opportunities to introduce concepts used later in the physical chemistry sequence. For instance, the discussion of Eyring's Transition State Theory allows the reintroduction of essential thermodynamic quantities: enthalpy, entropy, and Gibbs energy along with the equations that relate them. We must also reintroduce the concept of equilibrium. We find that little more than the presentations given in most general chemistry textbooks are necessary. The essential point is the assumption that the reaction rate for elementary reaction steps depends on an equilibrium established between the reactants and the transition state which in turn depends on the Gibbs energy of activation which can be separated into entropy and enthalpy factors. In order to connect TST to the Arrhenius theory we must describe how the enthalpy of activation is related to the activation energy. This requires a description of the internal energy which is described in terms of the standard models for molecular energetics: translation, rotation, and vibration. Only a qualitative description is necessary, but with it students obtain key insights regarding the distribution of energy within a sample of molecules. Granted, it is unsatisfying to present TST without statistical mechanics as we do, requiring that we quote the factor, "kT/h," without derivation. However, we feel that this has been a successful instructional presentation by observing our students' ability to apply and interpret the TST formulas correctly.

Another challenging but important topic is the expectation (or "average") value concept from quantum chemistry. The expectation value is difficult for students because it assumes the student can easily make the leap from a discrete formula for the mean value to a continuous mean formula cast as an integral. We help our students make this leap by constructing a concrete continuous average based on the Maxwell-Boltzmann speed distribution in the calculation of the average speed. Although a rather challenging topic in its own right, at least in the case of the MB distribution, the student can latch onto something they can visualize, namely the notion of a molecular speed and its distribution. Of course it is quite important to identify the traditional formula with a probability:

$$P(u) = \left(\frac{m}{2 \cdot \pi \cdot k \cdot T} \right)^{1/2} \cdot u^2 \cdot e^{-\frac{m \cdot u^2}{2 \cdot k \cdot T}}. \qquad \text{Eqn 5}$$

This leads eventually to:

$$\overline{u} = \int_0^\infty u \cdot P(u) \cdot du = \int_0^\infty u \cdot \left(\frac{m}{2 \cdot \pi \cdot k \cdot T} \right)^{1/2} \cdot u^2 \cdot e^{-\frac{m \cdot u^2}{2 \cdot k \cdot T}} \cdot du. \qquad \text{Eqn 6}$$

We show that this **continuous average** formula can be developed heuristically by artificially creating a discrete example from the continuous function. For a sample of nitrogen at room temperature, Eqn 5 is evaluated at twelve or thirteen speeds separated by an interval of 100 m/s. This pretty much spans the most significant range for nitrogen at 300 K. To make the problem even more concrete, a million molecule sample size is chosen. A rectangular box 1 ,/s wide is constructed at each of the "sampled" points so that the product of each evaluation of Eqn. 5 multiplied by one million gives the number of molecules (rounded to the one's place) represented by the area of each rectangular box. This gives us the data we need to calculate a discrete average. The logic for this problem is summarized in Figure 1. This discrete average gives a value that differs from the exact average speed by around 0.1 %. Our result is accurate in spite of the fact that we account for a small fraction of the original sample of molecules in the calculation, because our "sampling" of the probability distribution is sufficiently representative of the entire ensemble. By repeating the calculation using smaller intervals and a correspondingly larger number of rectangular boxes we show rapid convergence to the "exact" value. This construction convinces the typical student that the actual exact calculation is done with an infinite number of samplings of the probability distribution using an infinitesimal interval (with a smaller size box) in which the sum has been transformed into an integral.

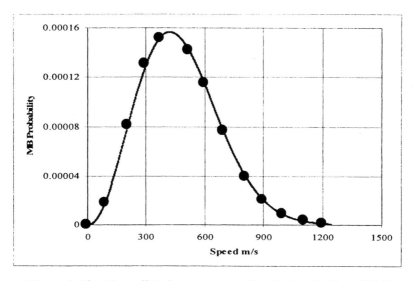

Figure 1. The Maxwell-Boltzmann speed distribution for N_2 at 300 K illustrating a calculation described in the text.

The ultimate goal is for students to realize that a continuous average of a quantity like $\bar{u}$ results from an integral of a product of that quantity times its probability distribution. The process of counting molecules provides a concrete application of a probability function. It can also be demonstrated that Eqn. 5 is normalized and then interpreted in terms of the number of molecules in the sample, another important quantum concept. This example can be completed as a directed homework exercise during the kinetic theory segment in the first semester and "resurrected" when quantum mechanics is taught. If the Maxwell-Boltzmann distribution example is followed by a problem utilizing one or more of the probability distributions encountered in the quantum chemistry segment of the course, the notion of an expectation value can be made more real for the student.

The Maxwell-Boltzmann distribution also provides further opportunities to introduce important quantum chemistry and mathematics topics. First, the derivation of the MB distribution, if one chooses to cover it, provides students with a look at a physics problem in three dimensions that also highlights (for review) the properties of the exponential function. This is a standard derivation found in most textbooks. Second, the velocity distribution bears a significant resemblance to the electron probability distribution for a 1s electron when displayed as a two-dimensional slice. The representation suggested here is a density of dots showing high density near the origin and decreasing in density in all radial directions. Although the 1s electron distribution is not Gaussian, it still exhibits the same spherical symmetry, and the opportunity to make a connection between these two models should not be overlooked. It is perfectly reasonable to present a 2-D slice of a 1s orbital as a density plot to compare with the corresponding plot in velocity space for the MB distribution. It is very important also to emphasize that the axis labels are x, y, and z in the case of the is orbital, but in the case of the velocity distribution the axis labels are u_x, u_y, and u_z. Third, continuing in this vein there is a direct parallel between the MB speed distribution and the radial distribution function for a spherically symmetric orbital. The MB speed distribution results from "integrating" the angular dependence found in the volume element for the MB velocity distribution. Because of spherical symmetry, this can be done geometrically by subtracting the volume of a sphere with radius u (r) from the volume of a sphere with radius u + du (r + dr) (5) neglecting higher powers of du (dr) in the algebra. The well-known volume element, $4 \pi u^2 du$ ($4 \pi r^2 dr$), results thus giving rise to the MB speed distribution in the case of kinetic theory and the radial distribution function in the case of the is orbital. This choice to reduce the number of variables allows us to more easily display the essential content of a three-dimensional problem. Plots of the two functions appear qualitatively similar and should be presented together, once again distinguishing the different labels for the horizontal coordinate. In a way, the interpretation is more intuitive for the radial distribution function being the variation of probability along any radial direction away from the nucleus. The speed distribution, also a probability, represents the likelihood that a sample of molecules will be found within a particular infinitesimal speed range. This is an opportunity to confront the apparent paradox in which the velocity distribution (probability distribution), which is maximum at the origin in one presentation, gives rise to a speed distribution (radial distribution function) that is zero at the origin. This results because the new volume element, $4 \pi u^2 du$, ($4 \pi r^2 dr$) is negligible near the origin.

Another way to derive the same result is to replace the Cartesian volume element, dx dy dz, with the volume element in spherical polar coordinates, $u^2 \sin\theta \, du \, d\theta \, d\phi$, and explicitly to integrate the angular parts to give the well-known result, 4π. Because of spherical symmetry this procedure provides a simpler visualization of the probability. Exactly why the volume element in spherical polar coordinates takes this form is a topic that one might choose to explain here making the presentation smoother once it comes up again during the quantum mechanics part of the course. The previous descriptions represent various presentations made over the past several years teaching the course. In a given year we would not attempt to present all of them, believing that it is unnecessary to derive every equation encountered.

Constraints on the Choice of Topical Order

The decision to change the order of topics depends on three considerations, at least: the needs of the students, the constraints of the department and institution, and the style and preferences of the instructor. At our institution, and at a large number of similar colleges and universities in the U.S., students seldom finish their bachelor's degree in the canonical four-year period. They often decide late in their careers to become chemistry majors and are hard-pressed to get ancillary courses like calculus and physics into their schedules before taking physical chemistry. It is no surprise that our student clientele is often marginally prepared for the physical chemistry course, and for most, the physical chemistry course is the last major hurdle to surmount before graduation. We have chosen to begin the course with what we believe to be more accessible material. We might have chosen to cover material in the more traditional order while glossing over much of the higher level mathematics, but that is not the goal of the approach described here. Instead we seek to provide a progressive introduction to physical chemistry that prepares our students for these more demanding topics and at the same time to make connections to chemistry courses previously taken.

At some institutions, however, external constraints prohibit changes in the order of topics, especially if Physical Chemistry I is a service course for other departments, most notably engineering. Further, there are no textbooks in which chemical kinetics is among the first topics, certainly among the most popular textbooks currently used. Our survey of physical chemistry teaching (*1*) identifies Atkins and dePaula (*13*), McQuarrie and Simon (*14*), and Engel and Reid (*15*) as the current top three most-used textbooks. With nearly 35% of the market, Atkins maintains a traditional, "Thermo First" topic ordering, while McQuarrie and Simon provides an elegant logical development beginning with quantum chemistry. Engle and Reid provide a "two volume" format that can be covered in either order allowing the instructor more flexibility to decide the sequence of topics.

Clearly the order of topics is an important consideration. Students' first impression of a course can be critical to capturing their interest and motivation. From the results of our survey it is apparent that there is a strong reluctance to change from the traditional order, "Thermodynamics, Kinetics, Quantum Mechanics, Statistical Mechanics" (TKQS). More than 25% of all respondents present physical chemistry in this exact order. In spite of the suggestion by Schwenz and Moore that quantum chemistry be taught first, only 20% of departments teach quantum first, with nearly two-thirds of all Physical Chemistry I courses beginning with thermodynamics (*1*). Only six of our 179 respondents to this survey say they began Physical Chemistry I with Chemical Kinetics, and all but one of those institutions were small, <5,000 students (In our survey intermediate size was considered to be from 5000 to 15000 students). From the general comments made by respondents in our survey it is clear that physical chemistry teachers are faced with a myriad of issues including student preparation, student demographics, calendar, appropriate textbook, etc. With so many considerations, an instructor's choice of the traditional order of topics may simply reflect an aversion to skipping around in the text. However, we feel it is most important that students succeed during the first segment of the physical chemistry course. Perhaps the arguments here will inspire others to experiment with this order of topics.

A Word on Assessment

We have been teaching physical chemistry with a "kinetics-first" orientation for 13 years. Over the course of this past decade we have also examined our students using the ACS Comprehensive Standardized Exam. Form 1995 (*16*) was used from 1996 through 2003 and Form 2002 (*17*) was used in 2004 and 2005. This comprehensive exam has been administered at the end of physical chemistry II covering quantum chemistry and spectroscopy. Both versions divide the 60 multiple choice questions into the three canonical areas, T, Q, and D, with the 1995 version assigning 20 questions each to the three areas. The 2002 version assigns 15 questions to the dynamics section and 25 to the quantum section. A few statistical mechanics questions are also scattered among these sections.

Our students have traditionally scored poorly on these exams compared with the national norms, which we have found to be at least partly correlated with SAT scores. In 1996 the average SAT score at Georgia Southern was barely 1000 while in 2006 it surpassed 1100. We do not try to "prep" our students for this examination and we tend not to give multiple choice hour exams in our physical chemistry courses. From 1996 through 2003, in eight classes of physical chemistry, 101 students have taken the 1995 exam. With 20 questions in a section, over this period; 2020 dynamics questions have been attempted and 805 have been answered correctly, a 39.9% rate. Over the same period, the same number of thermodynamics questions have been attempted and correctly answered at a 40.1 % rate while 40.9% of the quantum questions have been answered correctly. The difference in rates here is not statistically significant. Unfortunately, we did not teach any TKQS sections during this period and so we cannot compare these scores to a cohort who learned physical chemistry with the traditional ordering of topics. It is also somewhat strained to use student's performance on the thermodynamics section as a "control" comparison. Still, considering our students' background and foundation, we believe that introducing the material in the described order has not harmed their learning and has certainly removed some of the hurdles we previously encountered using a more traditional order.

In 2004 and 2005 we used the newer 2002 version of the Comprehensive exam. Unfortunately, student scores on these exams were slightly lower than scores on the previous exam although students appeared to do better on the dynamics questions than on the thermodynamics questions. The correct answer rate overall for 51 students in two classes was: D-38.0%, T-32.6%, and Q-38.7%. With factors such as different exam versions and a very small sample size (only 2 sections) it is impossible to draw definitive conclusions at this point. The D-T difference is barely significant, so it would be presumptuous to conclude that our order of topics is superior based on this one result. It is simply impossible to assess this later exam without the benefit of national norms for comparison. As always, we continue in our own observations and evaluations of this curriculum, and we would hope for others to try this approach so that a broader assessment could be conducted.

Summary

Teaching quantum mechanics first in physical chemistry gives the properly prepared and motivated student a presentation that best represents the most active areas of physical chemistry research. A thermodynamics first approach remains most consistent with the order found in most currently used textbooks avoiding the disorientation associated with skipping around in the textbook early in the course. However, we believe that a large segment of the chemistry major population is best served from the initial presentation: kinetic theory and kinetics. This presentation is more accessible mathematically, it is concrete and intuitive, and it represents a general area that provides a foundation for the study of dynamics, a field of chemistry that is at least as topical as quantum chemistry. The highly theoretical thermodynamics and quantum chemistry at best make a poor connection to the chemistry courses students might have previously taken, most specifically general chemistry and organic chemistry. Thermodynamics can bring in chemical reactions after a few weeks when thermochemistry is covered, but during the quantum chemistry segment much of the semester is consumed in the development of the formal structure before making connection to atoms, let alone to molecules, essentially the first example in which students might recognize "Chemistry." To cast students into the calculus of higher dimensions as is done after only a few weeks in both thermodynamics and a little longer in quantum chemistry leads to a highly stressful situation for the majority of chemistry majors.

A presentation that begins with kinetic theory and chemical kinetics begins with molecules and chemical reactions quickly and goes on to make connections both to general chemistry and to the organic chemistry sequence. If the introduction of material requires no more mathematics than is presented in calculus II, students have a better opportunity to succeed in their early exposure to physical chemistry perhaps coloring their attitude toward future topics. The delay of topics requiring higher mathematics allows the teacher more time to introduce calculus of higher dimensions as exercises integrated into homework or through mini-lectures during class-time. If laboratory is taught along with the physical chemistry I lecture, a teacher might use the propagation of error formula (*18*) as an intuitive application of multivariable calculus that is analogous to the total differential as an aid in introducing a "new" calculus topic.

According to our survey aimed to ascertain the current state of physical chemistry teaching, the order of topics is not considered an important method of reform. It is, however, interesting to note (*19*) that departments located at institutions with more than 15,000 students are most likely to choose to teach quantum chemistry first (11 out of 42, or 26%). Only 11 of 58 departments (19%) at schools with populations between 5,000 and 15,000, and only 12 of 75 departments (16%) at schools with populations below 5,000, offer quantum chemistry first (*1*). Further, 26% of all respondents continue to offer their curriculum in the traditional order TKQS. The choice of the order of topics in physical chemistry is certainly not a solution to the more difficult problem of inspiring students to see physical chemistry as a means to better understand the world, but if it can change student attitudes it might contribute to movement toward this goal. We hope that providing a successful experience early in the physical chemistry sequence through kinetic theory and kinetics will convert at least a few students into more open-minded thinkers who will not dread the rest of the physical chemistry course sequence and in the end take something useful from it.

References

1. A survey begun at the end of 2005 solicited 411ACS-accredited chemistry departments. The URL for this brief survey is: http://chemistryl.che.georgiasouthern.edu/jlobue/survey/PCsurveyQ.htm. This survey yielded a 44% response rate (179 respondents).
2. Sozbilir, M. *J. Chem. Ed.* **2004**, *81*, 573-578.
3. Moore, R. J.; Schwenz, R. W. *J. Chem. Ed.* **1992**, *69*, 1001-1002.
4. Comeford, L. *J. Chem. Ed.* **1997**, *74*, 392.
5. Moog, R. S.; Spencer, J. N.; Farrell, J. J., *Physical Chemistry: A Guided Inquiry: Thermodynamics;* Houghton Mifflin Co.: Boston, MA, 2004.
6. Heylin, M. *Chem. Eng. News,* July 24, 2006, p 45.
7. Borden, V. M. H. *Black Issues in Higher Education,* 2005, p 74.
8. Hahn, K. E.; Polik, W. F. *J. Chem. Ed.* **2004**, *81*, 567-572.
9. Laidler, K. A.; Meiser, J. H.; Sanctuary, B. C. *Physical Chemistry,* 4th Ed.; Houghton Mifflin Co.: Boston, MA, 2003; pp 17-33.
10. Alberty, R. A.; Silbey, R. J.; Bawendi, M. G. *Physical Chemistry,* 4th Ed.; John Wiley & Sons, Inc.: Hoboken, NJ, 2005; pp 623-631.
11. Levine, I. N. *Physical Chemistry,* 3rd Ed.; McGraw-Hill Book Co.: New York, NY; 2002; p 881.
12. Laidler, K. A.; Meiser, J. H.; Sanctuary, B. C. *Physical Chemistry,* 4th Ed.; Houghton Mifflin Co.: Boston, MA, 2003; pp 390-394.
13. Atkins, P.; de Paula, J. *Physical Chemistry,* 7th Ed.; W. H. Freeman and Co.: New York, NY; 2004.
14. McQuarrie, D. A.; Simon, J. D. *Physical Chemistry, a Molecular Approach,* University Science Books: Sausalito, CA; 1997.
15. Engel, T. and Reid, P. *Thermodynamics, Statistical Thermodynamics, & Kinetics;* Benjamin Cummings: San Francisco, CA; 2006.
16. Exams Institute of the American Chemical Society, Division of Chemical Education; *Physical Chemistry Examination (Comprehensive);* 1995; Clemson University, Clemson, SC.
17. Exams Institute of the American Chemical Society, Division of Chemical Education; *Physical Chemistry Examination (Comprehensive);* 2002; Clemson University, Clemson, SC.
18. Taylor, J. R. *In Introduction to Error Analysis,* 2nd Ed.; University Science Books: Sausolito, CA; 1997; pp 45-79.
19. Of the 179 respondents, only 175 indicated an institution size.

Appendix 1

Results of the Physical Chemistry Survey: *Below are excerpts of the survey data collected from physical chemistry faculty at ACS certified institutions.*

Distribution of institutions teaching Statistical Mechanics as part of a typical physical chemistry class:

During the first semester	16.8 %
During the second semester	54.2 %
Total	70.9 %

Commonly used physical chemistry textbooks:

Textbook (Author)	
Atkins and de Paula	34.9 %
McQuarrie and Simon	16.4 %
Engel and Reid	13.3 %
Laidler, Meiser, Sanctuary	8.7 %
Levine	8.2 %
Silbey, Alberty, Bawendi	8.2 %

Common order of topics:

Topic Order	
Thermo, Kinetics, Quantum, Stat Mech	26.1 %
Quantum, Stat Mech, Thermo, Kinetics	7.8 %
Thermo, Stat Mech, Kinetics, Quantum	5.0 %

First topic of discussion in physical chemistry courses (by institution size):

First Topic	Overall	Population < 5000	5000-15000	Population > 15000
Thermodynamics	71.8 %	62.7 %	75.9 %	64.3 %
Quantum	20.1 %	16.0 %	19.0 %	26.2 %
Kinetics	4.6 %	6.7 %	1.7 %	0.0 %
Statistical Mechanics	1.1 %			
Other	2.3 %			

Common physical chemistry course pre-requisites:

Analytical/Quantitative Analysis	27.8 %
Calculus II	58.3 %
Calculus III	29.4 %
Physics II – calculus based	59.4 %
Physics I & II – trig . Based	22.2 %

Appendix 2

Introduction	2 lectures

Relevant physics
Gases (Ideal gas law, Pressure, Units, and Conversion)

Kinetic Theory	4 lectures

Counting molecules, collision frequency, collision density
MB speed distribution and averages

Chemical Kinetics	6 lectures

Rate constant, rate law, concentration profile, experimental measurement, integrated rate laws, linear plots, half-lifes
Theory of the rate constant (activation energy, orientation factor, collision frequency factor, Transition State Theory)

Reaction mechanisms	4-6 lectures

Elementary steps, molecularity, rate determining step approximation, Steady State Approximation
Lindemann mechanism, Free radical chain mechanism, Enzyme kinetics, Surface chemistry

Thermodynamics	4 lectures

Gases, partial derivatives, triangle relation, inverse relation, total differential
Properties (Heat capacity, thermal expansion coefficient, isothermal expansivity coefficient)
Joule expt, Joule-Thomson expt, Relationship to intermolecular forces

First Law	4 lectures

Work, heat, ΔU, ΔH, Isothermal, isobaric, isochoric, adiabatic
Hess' law, calorimetry, T-dependence of enthalpy

Second Law	5 lectures

The entropy, Spontaneous vs non-spontaneous, Reversible and irreversible processes, Calculation of entropy changes (Isothermal, isobaric, isochoric, adiabatic), Phase changes at equilibrium, Trouton's rule, Calculation for irreversible processes

Third Law	1 lecture

Absolute zero, Calculation of third law entropies

Equilibrium	10-12 lectures

Gibbs energy, Helmholtz energy, Thermodynamic relationships
dU, dH, dG, dA
Derivations of thermodynamic quantities:
Chemical potential, chemical equilibrium (K_p, K_c, K_x), Phase equilibrium (1 component), Phase diagrams
Vapor pressure equation from Clapeyron eqn,
Phase equilibrium (2 component), liquid-vapor, solid-liquid, colligative properties, FP depression, osmotic pressure

Chapter 18

Fitting Physical Chemistry into a Crowded Curriculum: A Rigorous One-Semester Physical Chemistry Course with Laboratory

HollyAnn Harris

Department of Chemistry, Creighton University, Omaha, NE 68178

This chapter will describe how the physical chemistry curriculum at Creighton University was revised to include a one-semester overview of physical chemistry with laboratory, followed by elective courses in specific areas of physical chemistry. The course is preceded by a mathematics course designed specifically to prepare chemistry students for the mathematics encountered in a rigorous physical chemistry course.

The Creighton University Chemistry Department undertook full curriculum revision during the 2000 – 2001 academic year for both the ACS-certified major and the non-certified major. The revision was driven by the need (mandated by ACS) to include the equivalent of one semester of biochemistry in the major. We believed, at the time, that we did not have the overall expertise to add biochemistry units to existing courses and there was no room in the existing curriculum to simply add another course. The new biochemistry requirement forced a critical examination of a curriculum that had not changed in many years.

Prior to the revisions that occurred in 2001 the requirements for an ACS-certified degree in chemistry at Creighton University were very standard, one year each of general chemistry, organic chemistry, physical chemistry, and analytical chemistry (separated into quantitative and instrumental analysis), and one semester of advanced inorganic chemistry. Each of these courses had a laboratory specifically associated with it. In addition, we required an advanced elective in chemistry (most of the elective offerings were in the area of organic chemistry), a math course beyond Calculus II, and three credit hours of research. Including the pre-requisite courses in physics and calculus, the degree requirements summed to 61 credit hours making chemistry one of the largest majors, in terms of requirements, at Creighton University. Creighton is a Jesuit, Catholic university and has a relatively large core requirement (61 – 64 credit hours). Because there is very little overlap (8 credit hours) between the core requirements and the requirements for a major in chemistry, chemistry majors have very little room for electives or additional courses.

In order to add the required course in biochemistry and keep our major within reason with respect to the college requirements we made several changes in both analytical and physical chemistry. The most dramatic change was in the area of physical chemistry. Our new physical chemistry curriculum is loosely modeled after a curriculum that was used at Harvey Mudd College prior to 1988. The current curriculum consists of two lecture courses, one laboratory course and an elective (required only for the ACS certified degree). The lecture courses are *Math Concepts in Chemistry* (replaces the previous additional math course requirement) and *Introduction to Physical Chemistry* (a rigorous overview of the main topics in physical chemistry). The latter course is co-requisite with the laboratory course, *Physical Chemistry Laboratory* (a two-credit writing intensive course). Each course will be described in detail in the following sections.

In discussing physical chemistry curriculum revision we voiced many of the same concerns that are detailed in the New Traditions Physical Chemistry Curriculum Planning Session Report (*1*). Our new curriculum attempts to address specifically the concerns regarding math preparation, course content, active learning, writing skills, and appropriate utilization of the laboratory course to enhance learning.

Math Concepts in Chemistry

The *Math Concepts in Chemistry* course replaces the previous requirement of an additional math course. We have offered a course like this in the past but it was not required of all chemistry majors and was not a pre-requisite for physical chemistry. As such, students taking physical chemistry began the course with a variety of different backgrounds and skill levels in mathematics. The current course is required of all of our chemistry majors, whether or not they intend to complete the ACS degree requirements, and is a pre-requisite for the physical chemistry lecture. The goal of the course is to provide every student with the mathematical foundation necessary to grapple with the topics that will be covered in physical chemistry, as well as instrumental analysis and inorganic chemistry. The course is intended to be a math course, primarily, but because it is taught by a physical chemist the emphasis of the presentations is on applications of specific math techniques in chemistry.

A summary of the topics covered in the course is given in Table I. The prerequisite for the course is Calculus II. Given the number of topics covered in the course, we cannot present each topic in the mathematically rigorous way that our colleagues in the Math Department would prefer. However, our goal is not to train future mathematicians but rather to provide chemistry students with a level of familiarity with mathematical concepts that are useful in chemical applications. The chemical applications are emphasized in the examples used to present the math and in the problem sets, almost all of which highlight the uses of these mathematical concepts in chemistry.

The benefit of this course is that it provides all students taking the physical chemistry lecture course with the same mathematical foundation. In the physical chemistry lecture we can discuss the relationship between different thermodynamic functions without stopping to review partial derivatives. We can talk about the difference between work, heat, and energy without stopping to teach the difference between path functions and work functions. We can write

Table I. Topics Covered in *Math Concepts in Chemistry*

Topic	Approximate time spent
Functional forms and graphical representation	1.5 weeks
• Trigonometric functions	
• Exponentials and logarithms	
• Functions containing i	
Calculus review	2.5 weeks
• Differentiation, single- and multi-variable	
• Integration, single-and multi-variable	
• Min/Max theory	
Vectors, matrices, and determinants	3 weeks
Operator algebra	2 weeks
Differential equations	4 weeks
• Techniques for solving ODE's	
• Sequence and series solutions	
• Laplace transforms	
• Fourier transforms	
Introduction to symmetry and point groups	2 weeks

NOTE: The primary textbook for the course is Barrante, *Applied Mathematics for Physical Chemistry, 3E*, Pearson, Prentice Hall, 2004. Significant material is also taken from Mortimer, *Mathematics for Physical Chemistry, 3E*, Elsevier, 2005.

down the Schrödinger equation and know that the students understand what an operator is, in general, and are familiar with the Laplacian operator in particular.

The *Math Concepts* course meets two basic needs. It fulfills the ACS requirement for exposure to differential equations and linear algebra and it provides all students taking physical chemistry with the same math background. When the students encounter quantum mechanics in physical chemistry they can concentrate on the chemistry without having to learn the math simultaneously. An added benefit is that we can demand accountability of the students. Because the current *Math Concepts* course is a pre-requisite for physical chemistry we know that they have had previous exposure to the math because we are directly responsible for that exposure. In all but a very few instances the course is taken in the semester directly preceding the physical chemistry course so the mathematics should be fresh in the students' minds.

Introduction to Physical Chemistry

The *Introduction to Physical Chemistry* course is the centerpiece of the physical chemistry curriculum. It is named as such only because, by college rules, we needed to distinguish it from the previously offered Physical Chemistry I and Physical Chemistry II courses. We do believe, however, that it is aptly named because it provides an introduction to three of the four major areas of physical chemistry and our students are required to take an additional elective course covering one topic in greater detail.

The topics covered in this course are listed in Table II. This course is not a watered-down, non-mathematical treatment that is common in some one-semester courses. The mathematical rigor is retained because the math has been covered in the previous course, *Math Concepts in Chemistry*. By requiring the math concepts course we believe that we gain one-third to one-half of a semester worth of time in the physical chemistry course. Even taking into account the time saved by not "re-covering" math topics, we still needed to trim some content.

In choosing what topics to cover and what topics to cut we carefully considered an overall philosophy. We concluded that what makes physical chemistry different from the other major divisions of chemistry is that physical chemists are primarily concerned with constructing models that describe, and ultimately predict, chemical and physical behavior of matter. This conclusion is certainly open to argument and we don't intend to imply that content is unimportant. However, it is this philosophy that drove us to construct a course in which the content of the course focuses on *process* – the process of constructing models (primarily mathematical) that describe and predict chemical and physical behavior.

Table II. Topics Covered in Introduction to Physical Chemistry

Topic	Chapter	Approximate time spent
Kinetic theory and molecular motion	24	1.5 weeks
Gas laws, concentrating on non-ideal	1	1 week
Thermodynamic laws	2 – 5	3 weeks
Chemical equilibrium	6, 9	1.5 weeks
Quantum mechanics of atoms and molecules	11 – 15	5 weeks
Spectroscopy	16 – 18	3 weeks

NOTE: Chapter references are to the textbook used, Atkins and de Paula, *Physical Chemistry, 7E*, WH Freeman and Company, 2002.

Molecular emphasis

One topic that is conspicuously absent in the table above is statistical mechanics. While we do not cover statistical mechanics in the traditional sense (that topic is left to an elective course) we do emphasize, throughout the course, the molecular nature of matter. While the traditional coverage of thermodynamics is generally concerned with describing the behavior of the bulk sample, molecular interpretations (without resorting to a full ensemble explanation) can often provide very useful insight into bulk behavior and also reinforce the concept of model-building. By beginning the course with an investigation of molecular motion and a full development of the kinetic theory, students get a good introduction to individual molecular properties and distributions of molecular properties in a bulk sample. With this background we are able to use molecular interpretations to describe non-ideal behavior in both gas and condensed phases, as well as to justify enthalpies of reactions and enthalpies of mixing in addition to other thermodynamic phenomena. The statistical, molecular interpretation of entropy is also included in the course without a full statistical mechanics treatment.

Two other topics that are absent from the lecture course – kinetics and thermodynamics of condensed phases – are covered in the laboratory course, as well as in advanced elective courses.

Physical Chemistry Laboratory

Physical Chemistry Laboratory is a two-credit lab course meeting twice a week. The laboratory course complements the lecture course in two, disparate, ways. All of the experiments emphasize the theme of model-building and prediction of physical and/or chemical behavior. In addition to building on the theme of the lecture course, the laboratory course includes experiments that focus on material not covered in the lecture course, as well as experiments that illustrate topics covered in the lecture.

Specific experiments have changed from year to year as we continue to refine the course. Typical experiments include bomb calorimetry, kinetics of a reaction involving ions (allows coverage of basic kinetics as well as Debye-Hückel theory), a project designed to investigate the properties of ideal and non-ideal solutions, and spectroscopy. The solution project involves three to four experiments including solution calorimetry, viscosity, and construction of a liquid-vapor phase diagram. In this series of experiments we cover Raoult's law, viscosity, partial molar quantities, and thermodynamics of mixing – all topics that are not directly covered in the lecture course.

In addition to the content-specific goals of the laboratory, one very important goal is to introduce students to scientific writing and communication of experimental results. At least half of the experiments (including the solution project) culminate in a formal paper that is written in the style of an ACS journal article. The course is a college designated writing course which means that each paper must go through a graded draft and rewrite stage. In addition to the formal papers, the laboratory notebooks are graded critically for content and completeness. The notebook score comprises 30 % of a student's overall grade for the course and the formal papers comprise 50 – 60 % of the grade. The remaining 10 – 20 % of the grade is based on one or two oral presentations during the semester. Ideally each student will make one oral presentation (10 – 15 minutes) during the semester although we have not been consistent in this requirement to date. At the end of the semester each group of students chooses one experiment and makes a poster presentation at a department poster session. This poster session includes research posters as well as posters from the physical chemistry lab and is attended by the entire department as well as faculty from other science departments within our university.

A secondary goal of the laboratory course is the introduction of group work. For most of the experiments students work in small groups and must learn to co-ordinate their efforts. Most of the experiments are intentionally designed to "force" the students to cooperate, assign individual tasks and share data. Several of the papers are group writing projects, as well. This is a very different experience for our students as this is the first science course (and in many cases the only course of any kind) where teamwork is emphasized. This is not always a comfortable environment for our students but it is one that we, along with many others (1, 2), believe is important.

In constructing the particular experiments we have tried to emphasize the application of physical chemistry concepts to other fields of chemistry. Currently most of the experiments involve applications to organic chemistry but we are developing experiments that directly relate to inorganic (transition metal) chemistry and to biochemistry (2).

Elective Courses

All of our chemistry majors must take the *Math Concepts* course, the *Introduction to Physical Chemistry* lecture, and the *Physical Chemistry Laboratory*. In addition to these courses students who want to obtain the ACS-certified degree must also take a two-credit physical chemistry-based elective course. We average 25 majors in a graduating class and approximately 80 % of our majors obtain the ACS-certified degree. Ideally we would like to offer at least three elective courses each year so that students have a choice. At a minimum, two electives will always be offered.

Table III. Elective Courses

Course	Offered	Lecture / Laboratory
Statistical Mechanics	Fall, '04	Lecture
Kinetics	Spring, '05	Lecture
Chemical Applications of Spectroscopy	Spring, '05	Laboratory
Quantum Mechanics	Fall, '05	Lecture
Thermodynamics	Spring, '06	Lecture
Chemical Applications of Group Theory	Fall, '06	Lecture
Computational Chemistry	Spring, '07	Laboratory
Physical Chemistry of Macromolecules	Spring, '07	Lecture
Biophysical Chemistry*	Fall, '07	Lecture

NOTE: *Biophysical Chemistry* has been proposed but not yet approved. It is tentatively planned to be offered in Fall, '07.

The elective courses that we currently offer are listed in Table III. Two of the courses are laboratory courses. The offerings reflect the expertise of our current faculty and are subject to change, based on the interests of both the faculty and the students. The choice of which electives to offer in a given year is made by the faculty and students together. The faculty choose five or six possibilities from the list, based on teaching availability of particular faculty, and this list is given to the students in survey form, along with a course description, in the physical chemistry lecture course. The students fill out the survey rating their interest and the top two or three are chosen.

Because the electives are advanced courses in relatively narrow areas of physical chemistry it is easier to incorporate examples and applications from other fields of chemistry into the course. Often, the students have input into which applications will be covered. For example, in the Thermodynamics course that I just finished teaching, I wanted to spend some time investigating the benefits (if any), from both a thermodynamics and economics perspective, of ethanol-based additives in gasoline. We did that but we also spent more time on gas laws because the students wanted to investigate the recent claims by the tire industry that N_2 inflation was superior to air inflation. The resulting discussion and information (provided mostly by the students themselves) was fascinating and covered many topics including ideal and non-ideal gases and reactivity of mixed phases. I know that a similar student-led discussion occurred in the kinetics course.

Conspicuously missing from the list of electives (and anywhere in our curriculum) is a course on modern experimental physical chemistry and/or lasers in chemistry. We hope to correct this omission with a new hire in the next two years.

Assessment

We have used the ACS Physical Chemistry Comprehensive Exam (1995) as an assessment tool. As of this writing two groups of students have completed the core lecture and lab courses and taken the ACS exam. In discussing these exam results it is important to note that the exam was given to both groups (the two-semester group and the two one-semester groups) as an assessment activity. It did not count in the students' grades and they did not study or otherwise prepare for it.

The median total score on this exam for the students taking the one semester course is slightly higher (4 more correct answers) than the median score on the exam for students who took the previous, two-semester, version of physical chemistry. Of more interest to us are the scores on the individual components of the exam.

The exam (for those not familiar with it) is 60 questions divided into three categories – 20 questions each over Thermodynamics, Quantum Mechanics (including spectroscopy), and Dynamics (which includes one question on statistical mechanics). By looking at the section scores we can see that the improvement in total score comes from improvement in both the Thermodynamics and Dynamics sections. The median score for the Quantum Mechanics sections is the same for both groups of students. The largest improvement can be seen in the Thermodynamics section (median score 3 more correct for the one-semester group). A possible explanation for this is that the students in the two-semester group had a longer time lag between the coverage of thermodynamics in class and when the test was administered. However, the same explanation cannot be made for the improvement in the Dynamics sections. Students who took the two-semester sequence had a formal presentation of both kinetics and statistical mechanics in lecture and kinetics in lab. Students in the one-semester course had exposure to neither topic in lecture but did see kinetics in lab. We conclude from this admittedly small sample that the one-semester course is delivering (the same) content equivalent to the previous two-semester course.

In the laboratory course we are developing a rubric to assess the formal papers. The first and last papers will be scored according to this rubric to measure the degree to which writing skills have improved during the course. The poster presentations are assessed by chemistry faculty members outside of the physical chemistry division and those assessments are currently being evaluated by our department assessment committee.

We have had only two graduating classes complete the full series of physical chemistry courses, including the electives. In those two classes combined, 50 % of the students took more than one physical chemistry elective and two students in the class of 2005 took all three electives that were offered during that year. Based on the number of courses taken we conclude that most of our students are getting more physical chemistry with the new curriculum than they were with the traditional two-semester sequence.

One last piece of admittedly anecdotal evidence is that all five students who completed this curriculum, went on to graduate school in chemistry, and took a physical chemistry qualifying exam reported that they passed their exam. Unfortunately we do not have corresponding data for the previous curriculum so we cannot say whether or not the new curriculum is responsible for this performance. We can say, based on this limited sample, that the new curriculum certainly has not hurt these students.

Conclusions

Our physical chemistry curriculum revision is clearly a 'work in progress'. More work is needed so that the math course is more clearly and closely tied to the physical chemistry lecture course. The content of the lecture course needs to be refined and assessed. Finally, the laboratory experiments need to be modernized to more closely reflect current experimental physical chemistry.

However, we are gratified by the results so far. The preliminary assessment data supports our contention that we are not sacrificing rigor in the one-semester course. We are committed to the goals of developing teamwork and communication skills in the laboratory, as well as offering meaningful content and believe that we are well on our way to doing so.

The response, by both students and faculty, to the elective offerings has been much more positive than initially anticipated. Developing these advanced courses has energized the faculty and the students are very receptive to having input into what courses are offered and, in some cases, what topics are covered in the elective courses.

We are also pleased that the ACS Committee on Professional Training has recently endorsed the idea of one-semester "Foundations" courses (*3*). We hope that our curriculum, as we continue to refine and improve it, will serve as a model for how these foundational courses (in physical chemistry and other sub-disciplines of chemistry) might be constructed.

Acknowledgements

The author gratefully acknowledges the contributions of Dr. Kelly O. Sullivan, Pacific Northwest National Laboratories, for her enthusiasm, leadership, and ideas contributed during our curriculum revision discussions; Dr. Gerald Van Hecke, Harvey Mudd College, for his insight and endless willingness to discuss and debate issues surrounding physical chemistry education; and Drs. Mark Freitag and Robert Snipp, Creighton University, for their enthusiasm and willingness to develop 10 new courses in three years.

References

1. *Physical Chemistry Curriculum Planning Session Report*, http://newtraditions.chem.wisc.edu/PRBACK/pchem.htm; last accessed 4/28/06.
2. Zielinski, T.J.; Schwenz, R.W. *J. Chem. Ed.* **2001**, *78*, 1173-1174.
3. *Proposed Revision of the ACS Guidelines for Undergraduate Chemistry Programs*, American Chemical Society Committee on Professional Training, http://www.chemistry.org/portal/resources/ACS/ACSContent/education/CPT/ACS%20Proposed%20Guidelines%20Revision last accessed 3/31/06.

Indexes

Author Index

Subject Index

A

Ab initio calculations
  computers in laboratory, 125
  physical chemistry curriculum, 5
ACS. *See* American Chemical Society (ACS)
Activated processes, course content, 22–23
Active learning
  physical chemistry, 6
  teaching method, 101–102
Algebra, mathematics and physical chemistry, 83
American Chemical Society (ACS)
  comprehensive exam for assessment, 305–306
  guiding course content, 14
  philosophy of teaching, 280–281
  physical chemistry exam, 6
  physical chemistry survey results, 296
  Task Force on Chemical Education Research, 76
  See also Examinations
The American Journal of Physics
  teaching and learning quantum mechanics, 156
  See also Quantum mechanics
American system, physical chemistry courses, 44–45
Anharmonic oscillator
  calculation for Morse, 227*f*
  motion, 227*f*
  potential, kinetic and total energy, 226*f*
Approach, physical chemistry class, 31–32
Assessment
  Creighton University, 305–306

faculty, of material, 262–263
Georgia Southern, 292–293
student, of physical chemistry, 263, 265
Atomic force microscopy (AFM), laboratory experiments, 123, 126*t*
Atomic orbitals (AOs)
  modern concepts of, 91
  quantum chemistry, 89–93
Atomic spectroscopy, laboratory experiments, 134*t*
Atomistic simulations
  connecting macroscopic and molecular phenomena, 209–211
  See also Simulations
Aufbau principle, electrons in orbitals, 91–92
Authentic learning, *Virtual Substance*, 203–204
Availability, *Virtual Substance*, 205

B

Behaviors, studying quantum mechanics, 165–167
Biological systems, physical chemistry of, 129, 132*t*
Biology, impact on physical chemistry course, 47–48
Bohr model, physical chemistry education, 81
"Boss" model, relating to students, 34
Boundary surface, orbitals, 91
British Emergentism, philosophy, 69
British system, physical chemistry courses, 44–45
"Buddy" model, relating to students, 34
Buzzword, "nano" in physical chemistry, 2